Glencoe McGraw-Hill

Math Connects

Course 2

Skills Practice Workbook

To the Student This *Skills Practice Workbook* gives you additional examples and problems for the concept exercises in each lesson. The exercises are designed to aid your study of mathematics by reinforcing important mathematical skills needed to succeed in the everyday world. The materials are organized by chapter and lesson, with one *Skills Practice* worksheet for every lesson in *Glencoe Math Connects, Course 2*.

Always keep your workbook handy. Along with your textbook, daily homework, and class notes, the completed *Skills Practice Workbook* can help you review for quizzes and tests.

To the Teacher These worksheets are the same as those found in the Chapter Resource Masters for *Glencoe Math Connects, Course 2*. The answers to these worksheets are available at the end of each Chapter Resource Masters booklet as well as in your Teacher Wraparound Edition interleaf pages.

The McGraw·Hill Companies

Send all inquiries to:
Glencoe/McGraw-Hill
8787 Orion Place
Columbus, OH 43240

ISBN: 978-0-07-881053-4
MHID: 0-07-881053-1

Skills Practice Workbook, Course 2

Printed in the United States of America
2 3 4 5 6 7 8 9 10 047 14 13 12 11 10 09 08

CONTENTS

Lesson/Title		Page

NAME ______________________ DATE ____________ PERIOD _____

1-1 Skills Practice

A Plan for Problem Solving

Lesson 1-1

Answer these questions about the four-step problem-solving plan.

1. During which step do you ask if your answer makes sense?

2. During which step do you revise or make a new plan if your first plan doesn't work?

3. During which step do you select a strategy for solving the problem?

4. During which step do you ask yourself, "What do I need to find out?"

Choose one of the following to describe how you would plan to solve each problem. Do not solve the problems.

A. Use only one operation, such as addition or multiplication.

B. Use a combination of operations, such as division and addition.

C. Use a different strategy.

5. **MONEY** Julia opened a savings account with a deposit of $36. She then deposited $5 per week for one month. If she then withdrew $9.50, how much is left in her savings account?

6. In how many different patterns can 3 rose bushes, 2 sunflowers, and 5 tulip plants be planted in a garden?

7. Use the four-step plan to solve Exercise 5.

A. Understand

B. Plan

C. Solve

D. Check

NAME ______________________ DATE __________ PERIOD _____

1-2 Skills Practice

Powers and Exponents

Write each power as a product of the same factor.

1. 11^2

2. 3^4

3. 2^5

4. 9^3

5. 15^3

6. 4^3

7. 1^6

8. 17^4

9. 3^7

10. 8^6

Evaluate each expression.

11. 9^2

12. 8^2

13. 8^3

14. 2^4

15. 2^5

16. 6^3

17. 3^4

18. 3^5

19. 9^3

20. 11^2

21. 4^7

22. 12^3

23. 1^9

24. 10^4

25. 20^4

26. 2^6

Write each product in exponential form.

27. $12 \cdot 12$

28. $10 \cdot 10 \cdot 10$

29. $4 \cdot 4 \cdot 4 \cdot 4 \cdot 4$

30. $9 \cdot 9 \cdot 9 \cdot 9$

31. $15 \cdot 15 \cdot 15 \cdot 15 \cdot 15$

32. $6 \cdot 6 \cdot 6 \cdot 6 \cdot 6 \cdot 6 \cdot 6 \cdot 6$

NAME ______________________ DATE ____________ PERIOD _____

1-3 Skills Practice

Squares and Square Roots

Find the square of each number.

1. 3

2. 22

3. 25

4. 24

5. 35

6. 26

7. 37

8. 50

Find each square root.

9. $\sqrt{25}$

10. $\sqrt{100}$

11. $\sqrt{441}$

12. $\sqrt{900}$

13. $\sqrt{961}$

14. $\sqrt{784}$

15. $\sqrt{3,600}$

16. $\sqrt{1,936}$

17. What is the square of -37?

18. Find both square roots of 4,900.

19. Square 7.2.

20. Square 4.5.

Lesson 1-3

NAME ______________________ DATE __________ PERIOD _____

1-4 Skills Practice

Order of Operations

Evaluate each expression.

1. $9 - 3 + 4$

2. $8 + 6 - 5$

3. $12 \div 4 + 5$

4. $25 \times 2 - 7$

5. $36 \div 9(2)$

6. $6 + 3(7 - 2)$

7. $3 \times 6.2 + 5^2$

8. $(1 + 11)^2 \div 3$

9. $12 - (2 + 8)$

10. $15 - 24 \div 4 \cdot 2$

11. $(4 + 2) \cdot (7 + 4)$

12. $(3 \cdot 18) \div (2 \cdot 9)$

13. $24 \div 6 + 4^2$

14. $3 \times 8 - (9 - 7)^3$

15. $9 + (9 - 8 + 3)^4$

16. $3 \times 2^2 + 24 \div 8$

17. $(15 \div 3)^2 + 9 \div 3$

18. $(52 \div 4) + 5^3$

19. 26×10^3

20. 7.2×10^2

21. $5 \times 4^2 - 3 \times 2$

22. $24 \div 6 \div 2$

23. $13 - (6 - 5)^5$

24. $(8 - 3 \times 2) \times 6$

25. $(11 \cdot 4 - 10) \div 2$

26. $10 \div 2 \times (4 - 3)$

27. 1.82×10^5

28. $35 \div 7 \times 2 - 4$

29. $2^5 + 7(9 - 1)$

30. $12 + 16 \div (3 + 1)$

1-5 Skills Practice

Problem-Solving Investigation: Guess and Check

Solve each problem using the guess and check problem-solving strategy.

1. **SPORTS** Susan made 2-point baskets and 3-point baskets in her last basketball game. All together she scored 9 points. How many of each type of basket did she make?

2. **ENTERTAINMENT** Tickets to the local circus cost $3 for children and $5 for adults. There were three times as many children tickets sold as adult tickets. All together the circus made $700. How many children and how many adults bought tickets to the circus?

3. **NUMBERS** What are the next two numbers in the following sequence?

   5, 13, 37, 109, 325, ___, ___

4. **MONEY** Richard found $2.40 in change while cleaning his couch. He found the same number of quarters, dimes, and nickels. How many of each coin did he find?

NAME ______________________ DATE ____________ PERIOD _____

1-6 Skills Practice

Algebra: Variables and Expressions

Evaluate each expression if $w = 2$, $x = 3$, $y = 5$, and $z = 6$.

1. $2w$

2. $y + 5$

3. $9 - z$

4. $x + w$

5. $3 + 4z$

6. $6y - 5$

7. y^2

8. $y - x$

9. $\frac{z}{2}$

Evaluate each expression if $m = 3$, $n = 7$, and $p = 9$.

10. $m + n$

11. $12 - 3m$

12. $5p$

13. $3.3p$

14. $3.3p + 2$

15. $2p + 3.3$

16. $20 + 2n$

17. $20 - 2n$

18. $\frac{n}{7}$

19. n^2

20. $6m^2$

21. $\frac{p^2}{3}$

22. $1.1 + n$

23. $p - 8.1$

24. $3.6m$

25. $3n - 2m$

26. $3m - n$

27. $2.1n + p$

28. $\frac{m^2}{p}$

29. $\frac{2.5m + 2.5}{5}$

30. $\frac{(n + 2)^2}{3}$

NAME ______________________ DATE __________ PERIOD ____

1-7 Skills Practice

Algebra: Equations

Solve each equation mentally.

1. $a + 7 = 16$

2. $12 + x = 21$

3. $4d = 60$

4. $15 = \frac{u}{3}$

5. $\frac{b}{7} = 12$

6. $13 \cdot 3 = y$

7. $8 + r = 17$

8. $27 - 12 = m$

9. $h - 22 = 67$

10. $27 + 15 = n$

11. $36 + a = 96$

12. $99 \div d = 3$

13. $6t = 66$

14. $25 = y \div 4$

15. $b - 25 = 120$

16. $n \div 5 = 10$

17. $4y = 48$

18. $5t = 40$

19. $50 \cdot d = 150$

20. $w + 61 = 65$

21. $88 \div k = 2$

Graph the solution of each equation on a number line.

22. $v - 6 = 30$

23. $3a = 27$

24. $n + 7 = 14$

Define a variable. Write an equation and solve.

25. **BAKING** Judy wants to buy some cookies for her birthday party. Cookies come in packages of 6. If she is inviting 24 friends to her party, how many packages of cookies does she need to buy so that each of her friends can have one cookie each?

Lesson 1-7

1-8 Skills Practice

Algebra: Properties

Use the Distributive Property to write each expression as an equivalent expression. Then evaluate the expression.

1. $3(5 + 1)$

2. $(2 + 7)5$

3. $(10 + 2)7$

4. $2(9 - 8)$

5. $4(10 - 2)$

6. $6(13 + 4)$

Name the property shown by each statement.

7. $2 \times (3 \times 7) = (2 \times 3) \times 7$

8. $6 + 3 = 3 + 6$

9. $3(9 - 7) = 3(9) - 3(7)$

10. $18 \times 1 = 18$

11. $7 \times 2 = 2 \times 7$

12. $6 + (1 + 4) = (6 + 1) + 4$

13. $7 + 0 = 7$

14. $0 + 12 = 12$

15. $625 + 281 = 281 + 625$

16. $(12 \times 18) \times 5 = 12 \times (18 \times 5)$

17. $2(8 + 2) = 2(8) + 2(2)$

18. $(15 + 11) + 9 = 15 + (11 + 9)$

19. $(6 + r) + s = 6 + (r + s)$

20. $(4 \times 8) \times a = 4 \times (8 \times a)$

21. $p \times 1 = p$

22. $a + 5 = 5 + a$

23. $y \times 3 = 3 \times y$

24. $b + 0 = b$

25. $(x + y) + z = x + (y + z)$

26. $6(200 + 50) = 6(200) + 6(50)$

NAME ______________________ DATE ____________ PERIOD _____

1-9 Skills Practice

Algebra: Arithmetic Sequences

Describe the relationship between the terms in each arithmetic sequence.

1. 3, 6, 9, 12...

2. 1, 3, 5, 7, ...

3. 1, 2, 3, 4, ...

4. 0, 7, 14, 21, ...

5. 2, 5, 8, 11, ...

6. 5, 10, 15, 20, ...

7. 0.3, 0.6, 0.9, 1.2, ...

8. 1, 10, 19, 28, ...

9. 12, 18, 24, 30, ...

10. 0.5, 2.5, 4.5, 6.5, ...

11. 3, 7, 11, 15, ...

12. 0, 4.5, 9, 13.5, ...

13. 11, 22, 33, 44, ...

14. 11, 19, 27, 35, ...

Give the next three terms in each sequence.

15. 3, 6, 9, 12, ...

16. 18, 21, 24, 27, ...

17. 7, 10, 13, 16, ...

18. 4, 8, 12, 16, ...

19. 0, 7, 14, 21, ...

20. 7, 12, 17, 22, ...

21. 5, 7, 9, 11, ...

22. 5, 15, 25, 35, ...

23. 21, 42, 63, 84, ...

24. 1.1, 2.2, 3.3, 4.4, ...

25. 0.5, 1.0, 1.5, 2.0, ...

26. 1.7, 1.9, 2.1, 2.3, ...

27. 0.5, 1.5, 2.5, 3.5, ...

28. 0.1, 0.2, 0.3, 0.4, ...

NAME ______________________ DATE ____________ PERIOD ______

1-10 Skills Practice

Algebra: Equations and Functions

Copy and complete each function table. Identify the domain and range.

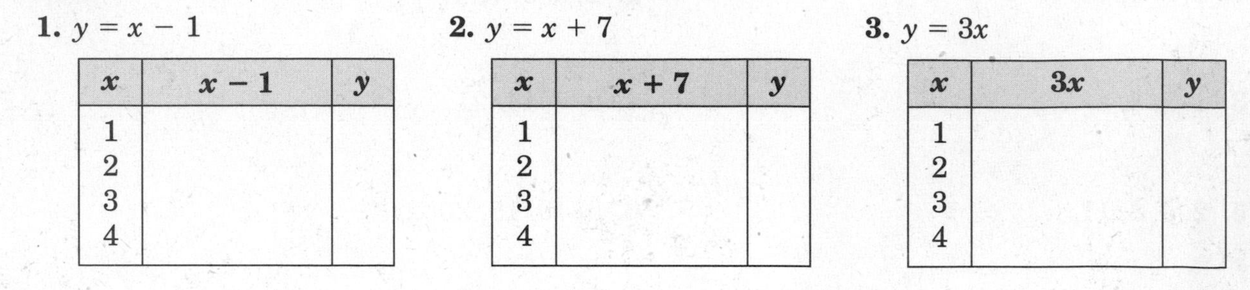

1. $y = x - 1$

x	$x - 1$	y
1		
2		
3		
4		

2. $y = x + 7$

x	$x + 7$	y
1		
2		
3		
4		

3. $y = 3x$

x	$3x$	y
1		
2		
3		
4		

4. $y = 4x$

x	$4x$	y
2		
3		
4		
5		

5. $y = x - 0.5$

x	$x - 0.5$	y
1		
2		
3		
4		

6. $y = 10x$

x	$10x$	y
0		
1		
2		
3		

Solve each word problem.

For Exercises 7 and 8, use the following information.

TRAVEL **For every gallon of gas, a car can travel 30 miles.**

7. Write an equation using two variables to show the relationship between the distance the car travels and the gallons of gas it uses.

8. If a car had 8 gallons of gas left in its tank, how many miles can it travel before the tank runs out?

For Exercises 9 and 10, use the following information.

FARMING **Every row of corn in Mr. Jones' garden has 5 cornstalks.**

9. Write an equation using two variables to show the relationship between the number of rows and the number of cornstalks.

10. If Mr. Jones has 7 rows of corn, how many cornstalks will he need to harvest?

NAME ______________________ DATE ____________ PERIOD _____

2-1 Skills Practice

Integers and Absolute Value

Write an integer for each situation.

1. 15°C below 0

2. a profit of $27

3. 2010 A.D.

4. average attendance is down 38 people

5. 376 feet above sea level

6. a withdrawal of $200

7. 3 points lost

8. a bonus of $150

9. a deposit of $41

10. 240 B.C.

11. a wage increase of $120

12. 60 feet below sea level

Evaluate each expression.

13. $|-1|$

14. $|9|$

15. $|23|$

16. $|-107|$

17. $|-45|$

18. $|19|$

19. $|0|$

20. $|6| - |-2|$

21. $|-8| + |4|$

22. $|-12| - |12|$

Graph each set of integers on a number line.

23. {0, 2, −3}

−4 −3 −2 −1 0 1 2 3 4

24. {−4, −1, 3}

−4 −3 −2 −1 0 1 2 3 4

2-2 Skills Practice

Comparing and Ordering Integers

Replace each ● with < or > to make a true sentence.

1. -15 ● -16

2. -8 ● -7

3. 0 ● -2

4. -2 ● -5

5. -25 ● 3

6. -14 ● $|-20|$

7. $|-4|$ ● 3

8. $|-6|$ ● $|-7|$

9. $|-7|$ ● $|2|$

10. -8 ● $|-9|$

Determine whether each sentence is *true* or *false*. If *false*, change one number to make the sentence true.

11. $-7 < 3$

12. $2 > 0$

13. $-20 < -22$

14. $12 < 15$

15. $3 > |-5|$

16. $|-2| < -3$

17. $|8| < |-10|$

18. $|-11| = 11$

19. $-4 < 4$

20. $|-9| < |-10|$

Order the integers from least to greatest.

21. 12, −6, 20, −47, −11

22. 9, −6, 0, −4, 17, −11

Order the integers from greatest to least.

23. −40, 65, −7, 24, −6, 15

24. $|-13|$, 0, 7, −8, −5, $|2|$

2-3 Skills Practice

The Coordinate Plane

Name the ordered pair for each point graphed at the right. Then identify the quadrant in which each point lies.

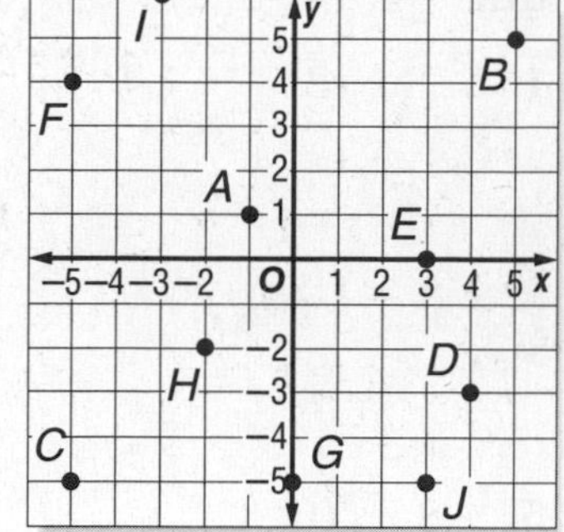

1. *A*

2. *B*

3. *C*

4. *D*

5. *E*

6. *F*

7. *G*

8. *H*

9. *I*

10. *J*

Graph and label each point on the coordinate plane.

11. $N(-1, 3)$

12. $V(2, -4)$

13. $C(4, 0)$

14. $P(-6, 2)$

15. $M(-5, 0)$

16. $K(-1, 5)$

17. $I(-3, -3)$

18. $A(5, -3)$

19. $D(0, -5)$

Name the ordered pair for each point on the city map at the right.

20. City Hall

21. Theater

22. Gas Station

23. Grocery

Grocery

Gas Station

City Hall

Theater

NAME ______________________ DATE ____________ PERIOD _____

2-4 Skills Practice

Adding Integers

Add.

1. $5 + (-8)$

2. $-3 + 3$

3. $-3 + (-8)$

4. $-7 + (-7)$

5. $-8 + 10$

6. $-7 + 13$

7. $15 + (-10)$

8. $-11 + (-12)$

9. $25 + (-12)$

10. $-14 + (-13)$

11. $14 + (-27)$

12. $-28 + 16$

Evaluate each expression if $a = -8$, $b = 12$, and $c = -4$.

13. $5 + a$

14. $b + (-9)$

15. $c + (-5)$

16. $a + b$

17. $a + 0$

18. $b + c$

19. $-12 + b$

20. $a + (-7)$

21. $21 + c$

22. $a + c$

NAME ______________________ DATE __________ PERIOD ____

2-5 Skills Practice

Subtracting Integers

Subtract.

1. $5 - 2$

2. $6 - (-7)$

3. $-3 - 2$

4. $8 - 13$

5. $-7 - (-7)$

6. $6 - 12$

7. $15 - (-7)$

8. $-15 - 6$

9. $-3 - 8$

10. $-10 - 12$

11. $13 - (-12)$

12. $14 - (-22)$

13. $10 - (-20)$

14. $-16 - 14$

15. $-25 - 25$

16. $6 - (-31)$

17. $-18 - (-40)$

18. $15 - (-61)$

Evaluate each expression if $r = -4$, $s = 10$, and $t = -7$.

19. $r - 7$

20. $t - s$

21. $s - (-8)$

22. $t - r$

23. $s - t$

24. $r - s$

Lesson 2-5

NAME ______________________ DATE ____________ PERIOD _____

2-6 Skills Practice

Multiplying Integers

Multiply.

1. $-4(6)$

2. $-2(-8)$

3. $12(-4)$

4. $-6(5)$

5. $-10(-9)$

6. $-(5)^2$

7. $(-5)^2$

8. $-30(5)$

9. $20(-6)$

10. $-14(-6)$

11. $(-13)^2$

12. $-7(15)$

ALGEBRA Simplify each expression.

13. $-3(4y)$

14. $7(-3x)$

15. $7(5g)$

16. $7(7w)$

17. $3(-3y)$

18. $-2(-10h)$

ALGEBRA Evaluate each expression if $g = -5$, $h = -3$, and $k = 4$.

19. $-3g$

20. $5h$

21. $7gk$

22. $-2gh$

23. $-10h$

24. $-2h^2$

NAME ______________________ DATE ____________ PERIOD _____

2-7 Skills Practice

Problem-Solving Investigation: Look for a Pattern

Look for a pattern to solve the problem.

1. **NUMBERS** What are the next two numbers in the pattern listed below?
   7, 21, 63, 189, ...

2. **POPULATION** The Springfield Zoo is breeding gorillas. They have 3 gorillas, which can mate and give birth. After the first year there are 7 gorillas. After the second year there are 11 gorillas. If the gorillas continue to increase at the same rate, how long will it take for the Springfield Zoo to have 35 gorillas?

3. **ALGEBRA** Read the table below to find a pattern relating x and y. Then write an equation to describe the pattern in general.

x	y
1	5
2	8
3	11
4	14
5	17

4. **SAVINGS** Maria receives $50 for her birthday. She decides to put the money into a bank account and start saving her money from babysitting in order to buy a television that costs $200. After the first week she has $74. After the second week, she has $98. After the third week she has $122. How many weeks will she have to save at the same rate in order to buy the television?

Lesson 2-7

NAME ______________________ DATE ____________ PERIOD _____

2-8 Skills Practice

Dividing Integers

Divide.

1. $-15 \div 3$

2. $-24 \div (-8)$

3. $22 \div (-2)$

4. $-49 \div (-7)$

5. $-8 \div (-8)$

6. $\frac{36}{-4}$

7. $225 \div (-15)$

8. $\frac{0}{-9}$

9. $-38 \div 2$

10. $\frac{64}{4}$

11. $-500 \div (-50)$

12. $-189 \div (-21)$

ALGEBRA Evaluate each expression if $m = -32$, $n = 2$, and $p = -8$.

13. $m \div n$

14. $p \div 4$

15. $p^2 \div m$

16. $m \div p$

17. $\frac{-p}{n}$

18. $p \div n^2$

19. $\frac{p^2}{n^2}$

20. $\frac{18 - n}{p}$

21. $m \div (np)$

22. $\frac{m}{p} + n$

NAME ______________________ DATE ____________ PERIOD _____

3-1 Skills Practice

Writing Expressions and Equations

Write each phrase as an algebraic expression.

1. b plus 1

2. three more than x

3. twelve minus y

4. seven less than n

5. five years younger than Jessica

6. a number less eleven

7. four increased by a

8. eight dollars more than m

9. the product of c and 10

10. twice as many days

11. three times as many soft drinks

12. t multiplied by 14

13. Emily's age divided by 3

14. 24 divided by some number

15. a number divided by 2

16. the quotient of -15 and w

Write each sentence as an algebraic equation.

17. A number plus three is 9.

18. The sum of x and 2 is 10.

19. Four cents more than the price is 93¢.

20. Fifteen minus y is 7.

21. A number decreased by 5 is 12.

22. Five dollars less than Yumi's pay is $124.

23. A number times four is 20.

24. Twice the number of cars is 40.

25. The product of z and 6 is 54.

26. A number divided by 6 is 12.

27. 72 divided by y is -9.

28. 175 students separated into n classes is 25.

29. One more than twice as many CDs is 17.

30. Four less than three times a number is 14.

NAME ______________________ DATE ____________ PERIOD _____

3-2 Skills Practice

Solving Addition and Subtraction Equations

Solve each equation. Check your solution.

1. $x + 2 = 8$

2. $y + 7 = 9$

3. $a + 5 = 12$

4. $16 = n + 6$

5. $q + 10 = 22$

6. $m + 9 = 17$

7. $b - 4 = 9$

8. $8 = c - 4$

9. $11 = t - 7$

10. $d - 10 = 8$

11. $x - 11 = 9$

12. $2 = z - 14$

13. $72 = 24 + w$

14. $86 + y = 99$

15. $6 + y = -8$

16. $-5 = m + 11$

17. $n + 3.5 = 6.7$

18. $x + 1.6 = 0.8$

19. $98 = t - 18$

20. $12 = g - 56$

21. $x - 18 = -2$

22. $p - 11 = -5$

23. $a - 1.5 = 4.2$

24. $7.4 = n - 2.6$

NAME ______________________ DATE ____________ PERIOD ______

3-3 Skills Practice

Solving Multiplication Equations

Solve each equation. Check your solution.

1. $4c = 16$	**2.** $10x = 50$	**3.** $42 = 6s$
4. $9c = 45$	**5.** $49 = 7y$	**6.** $11t = 44$
7. $15a = 60$	**8.** $72 = 12c$	**9.** $18x = 162$
10. $14d = 154$	**11.** $24z = 288$	**12.** $16v = 256$
13. $-5b = 40$	**14.** $32 = -2f$	**15.** $-9x = -63$
16. $4g = -52$	**17.** $-5x = -85$	**18.** $-63 = 7a$
19. $0.6m = 1.8$	**20.** $1.5z = 6$	**21.** $0.6q = 3.6$
22. $1.8a = 0.9$	**23.** $1.2r = 4.8$	**24.** $2.4 = 0.2t$

3-4 Skills Practice

Problem-Solving Investigation: Work Backward

Solve. Use the work backward strategy

1. **GOVERNMENT** There are 99 members in the Ohio House of Representatives. All of them were present when a vote was taken on a piece of legislation. If 6 of them did not vote, and 13 more voted "yes" than voted "no", how many "no" votes were there?

2. **MONEY** Jessie and Amar eat lunch at a restaurant and their bill is $21.65. Amar gives the cashier a coupon for $6 off their bill, and also hands the cashier two bills. If he receives $4.35 in change, what were the denominations of the two bills he gave the cashier?

3. **AGE** Justine is 13 years younger than her uncle Stewart. Stewart is 18 years older than Justine's sister, Julia. Julia's mother is 8 year older than Stewart, and 28 years older than her youngest child, Jared. If Jared is 12 years old, how old is Justine?

4. **NUMBER THEORY** A number is divided by 6. Then 7 is added to the divisor. After dividing by 4, the result is 4. What is the number?

5. **COMPACT DISCS** Carmella borrowed half as many CDs from the library as her friend Ariel. Ariel borrowed 2 more than Juan, but four less than Sierra. Sierra borrowed 12 CDs. How many did each person borrow?

6. **TIME** Ashish needs to leave for the bus stop 15 minutes earlier than his friend Rami. Rami leaves five minutes later than Susan, but 10 minutes earlier than Raphael. If Raphael leaves for the bus stop at 8:15, what time does Ashish need to leave?

NAME ______________________ DATE ____________ PERIOD _____

3-5

Skills Practice

Solving Two-Step Equations

Solve each equation. Check your solution.

1. $2x + 1 = 9$

2. $5b + 2 = 17$

3. $3w + 5 = 23$

4. $8n + 1 = 25$

5. $4t - 2 = 14$

6. $7k - 3 = 32$

7. $8x - 1 = 63$

8. $2x - 5 = 15$

9. $3 + 6v = 45$

10. $9 + 4b = 17$

11. $2p + 14 = 0$

12. $3y + 10 = -2$

13. $3w + 5 = 2$

14. $8x + 7 = -9$

15. $5d - 1 = -11$

16. $4d - 35 = -3$

17. $11x - 24 = -2$

18. $15a - 54 = -9$

19. $3g - 49 = -7$

20. $-2x - 4 = 8$

21. $-9d - 1 = 17$

22. $-4f + 1 = 13$

23. $-5b + 24 = -1$

24. $-6x + 4 = -2$

Lesson 3-5

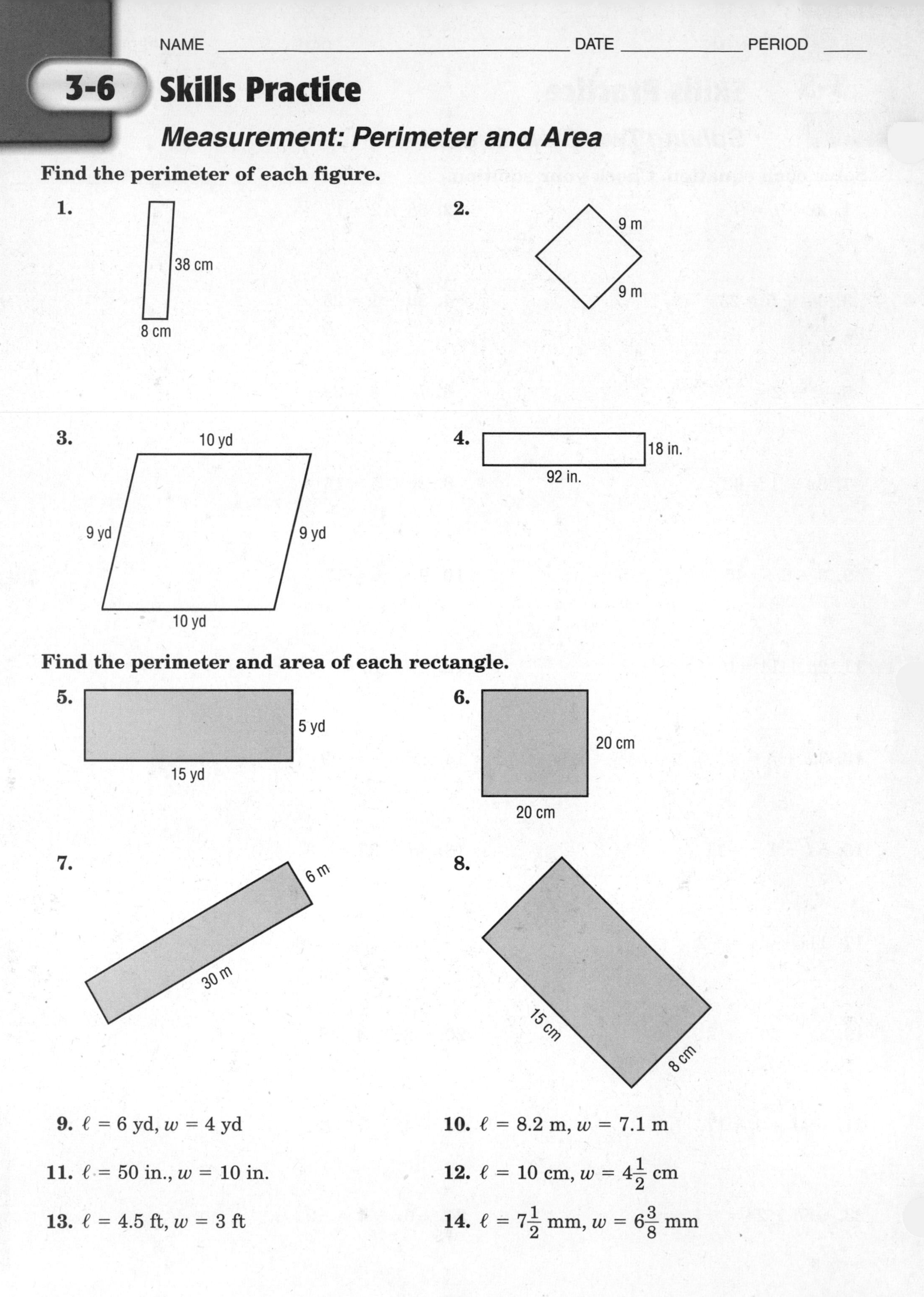

NAME ______________________ DATE ____________ PERIOD _____

3-6 Skills Practice

Measurement: Perimeter and Area

Find the perimeter of each figure.

1.

38 cm
8 cm

2.

9 m
9 m

3.

10 yd
9 yd
9 yd
10 yd

4.

18 in.
92 in.

Find the perimeter and area of each rectangle.

5.

5 yd
15 yd

6.

20 cm
20 cm

7.

6 m
30 m

8.

15 cm
8 cm

9. $\ell = 6$ yd, $w = 4$ yd

10. $\ell = 8.2$ m, $w = 7.1$ m

11. $\ell = 50$ in., $w = 10$ in.

12. $\ell = 10$ cm, $w = 4\frac{1}{2}$ cm

13. $\ell = 4.5$ ft, $w = 3$ ft

14. $\ell = 7\frac{1}{2}$ mm, $w = 6\frac{3}{8}$ mm

NAME ______________________ DATE ____________ PERIOD _____

3-7 Skills Practice

Functions and Graphs

Copy and complete each function table.

1. $y = x - 1$

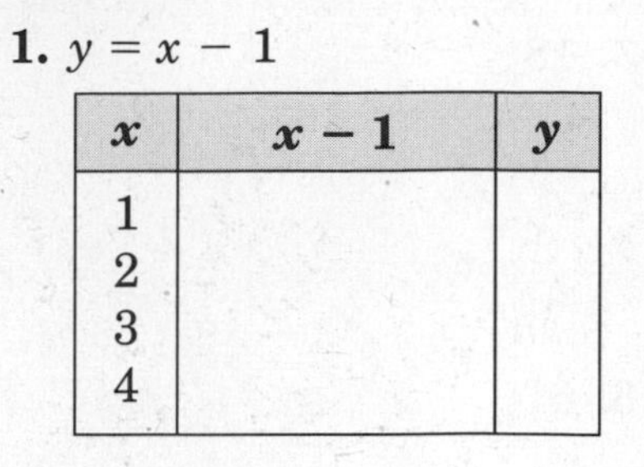

x	$x - 1$	y
1		
2		
3		
4		

2. $y = x + 7$

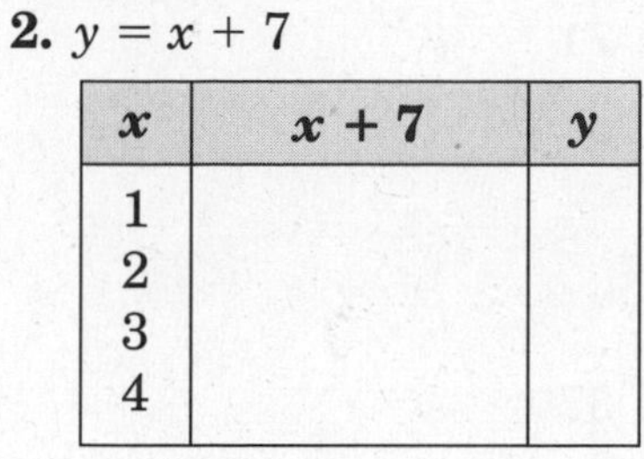

x	$x + 7$	y
1		
2		
3		
4		

3. $y = 3x$

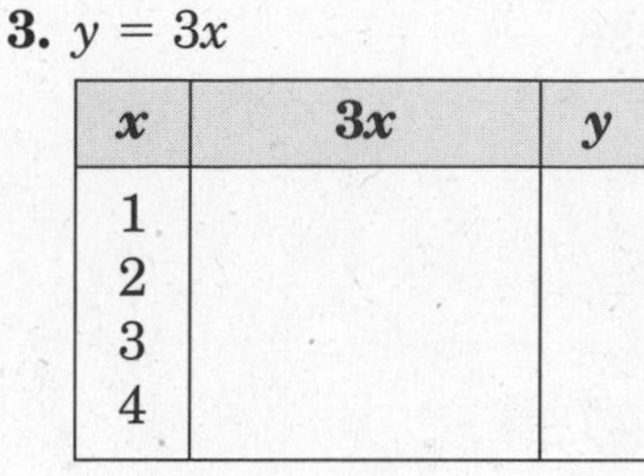

x	$3x$	y
1		
2		
3		
4		

4. $y = -4x$

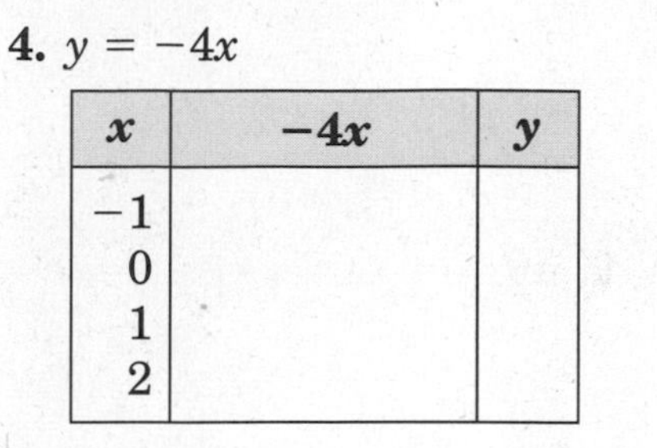

x	$-4x$	y
−1		
0		
1		
2		

5. $y = 3x + 1$

x	$3x + 1$	y
−1		
0		
1		
2		

6. $y = -2x + 3$

x	$-2x + 3$	y
−1		
0		
1		
2		

Graph each equation.

7. $y = x - 2$

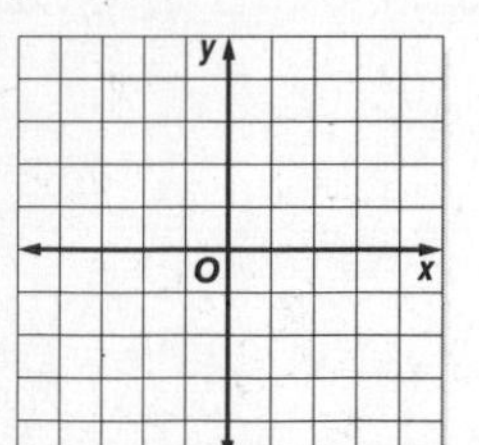

8. $y = x + 4$

9. $y = -3x$

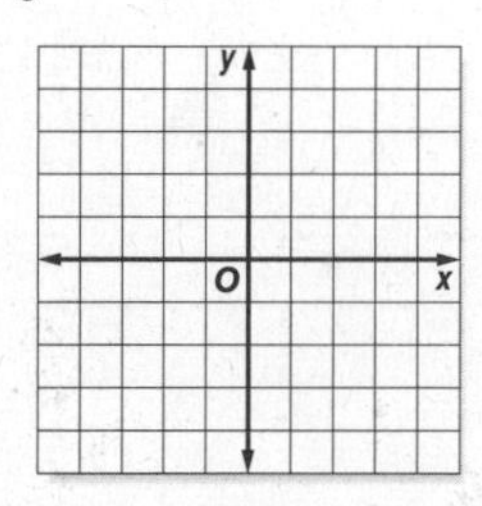

10. $y = 2x$

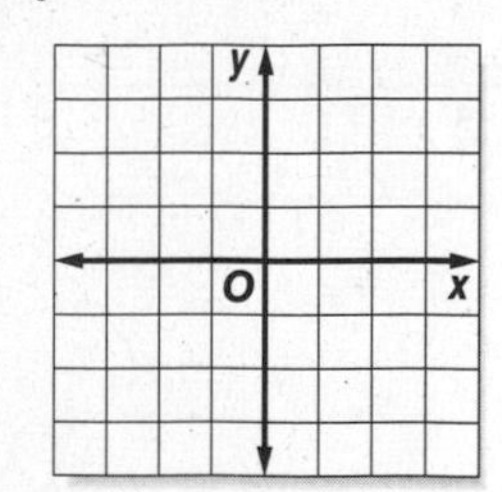

11. $y = 2x + 2$

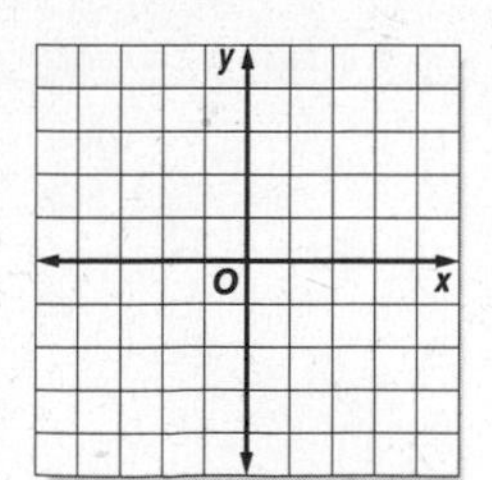

12. $y = 3x - 2$

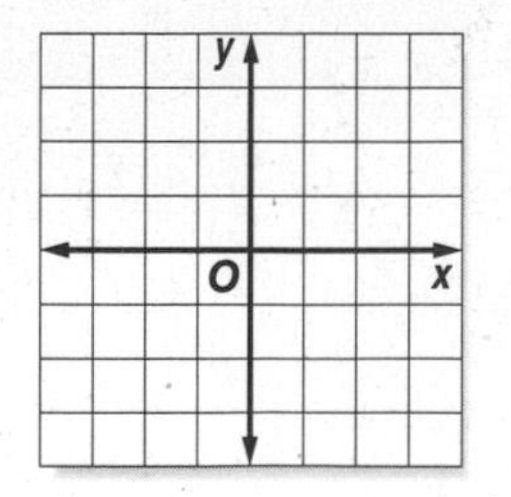

13. $y = 0.75x$

14. $y = 0.5x + 1$

15. $y = 2x - 0.5$

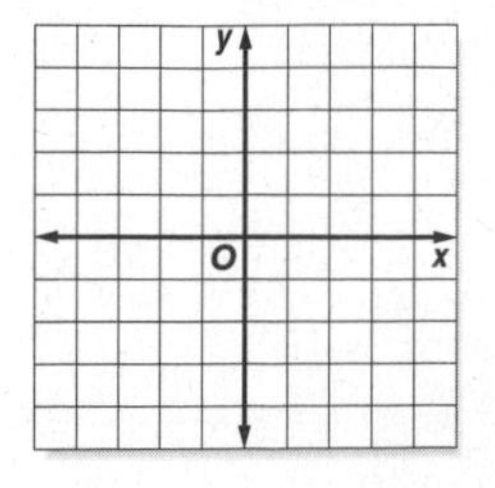

NAME _______________________________ DATE ______________ PERIOD _____

4-1 Skills Practice

Prime Factorization

Determine whether each number is *prime* or *composite*.

1. 36 **2.** 71 **3.** 18

4. 27 **5.** 37 **6.** 61

7. 32 **8.** 21 **9.** 40

Find the prime factorization of each number.

10. 425 **11.** 82 **12.** 93

13. 142 **14.** 45 **15.** 56

16. 63 **17.** 236 **18.** 12

19. 110 **20.** 46 **21.** 84

NAME ______________________ DATE ____________ PERIOD _____

4-2 Skills Practice

Greatest Common Factor

Find the GCF of each set of numbers.

1. 14, 20

2. 16, 42

3. 8, 18

4. 24, 36

5. 72, 22

6. 77, 15

7. 32, 80

8. 90, 120

9. 45, 30

10. 12, 62

11. 15, 27

12. 21, 28

13. 12, 20, 26

14. 15, 20, 25

15. 60, 72, 36

16. 32, 48, 64

17. 36, 48, 30

18. 28, 56, 42

19. 80, 110, 90

20. 9, 25, 49

Find the GCF of each set of algebraic expressions.

21. $21ab$, $14b$

22. $20a^2$, $36a$

23. $15ab$, $5b^2$

24. $35a^2$, $85ab$

25. Find the GCF of $2^3 \times 3^2 \times 5$ and $2^2 \times 3 \times 5^2$.

NAME ______________________ DATE ____________ PERIOD _____

4-3 Skills Practice

Problem-Solving Investigation: Make An Organized List

Solve by making an organized list.

1. **BAKING** Virginia and Robert have 1 dozen of each of the following types of cookies: chocolate chip, oatmeal raisin, snickerdoodles, and shortbread. If they want to divide the cookies into packages of two dozen, with one dozen of each of two types of cookie per package, how many different ways can they group the cookies?

2. **NUMBER THEORY** How many different two-digit numbers can be made using the digits 2, 9, 6, and 3?

3. **FOOD** Takanae is ordering lunch at a deli and is trying to decide what she would like on her sandwich. She has her choice of turkey, ham, or roast beef and a choice of cheddar, swiss, or muenster cheese. How many combinations of sandwich could she choose assuming that each sandwich has one type of meat and one type of cheese?

4. **TELEPHONES** How many phone numbers are possible for one area code if the first four numbers are 202-1, in that order, and the last three numbers are 1-7-8 in any order?

5. **CLOTHES** Sheila has four different shirts and two skirts with her on a business trip. How many different outfits can she create?

6. **SPORTS** Juan and Andrew are planning the schedule for a softball tournament. If there are 6 teams, how many different pairings could they make for the first tournament game?

NAME ______________________ DATE ____________ PERIOD _____

4-4 Skills Practice

Simplifying Fractions

Write each fraction in simplest form.

1. $\frac{49}{70}$

2. $\frac{5}{30}$

3. $\frac{6}{14}$

4. $\frac{14}{28}$

5. $\frac{72}{72}$

6. $\frac{18}{21}$

7. $\frac{45}{75}$

8. $\frac{50}{200}$

9. $\frac{32}{50}$

10. $\frac{56}{64}$

11. $\frac{14}{35}$

12. $\frac{39}{45}$

13. $\frac{48}{66}$

14. $\frac{42}{45}$

15. $\frac{78}{130}$

Write two fractions that are equivalent to each fraction.

16. $\frac{3}{4}$

17. $\frac{7}{9}$

18. $\frac{7}{11}$

19. $\frac{14}{17}$

20. $\frac{21}{23}$

21. $\frac{11}{17}$

NAME ______________________________ DATE ____________ PERIOD _____

4-5 Skills Practice

Fractions and Decimals

Write each repeating decimal using bar notation.

1. 0.7353535...

2. 0.424242...

3. 5.126126126...

Write each fraction or mixed number as a decimal. Use bar notation if the decimal is a repeating decimal.

4. $\frac{3}{5}$

5. $\frac{19}{20}$

6. $3\frac{4}{5}$

7. $\frac{23}{50}$

8. $1\frac{5}{8}$

9. $\frac{19}{25}$

10. $4\frac{17}{37}$

11. $5\frac{3}{11}$

12. $\frac{17}{24}$

13. $6\frac{7}{32}$

14. $7\frac{9}{22}$

15. $1\frac{17}{48}$

Write each decimal as a fraction in simplest form.

16. 0.8

17. 0.52

18. 0.92

19. 0.48

20. 0.86

21. 0.76

NAME ______________________ DATE ____________ PERIOD _____

4-6 Skills Practice

Fractions and Percents

Write each ratio as a percent.

1. 26 out of 100

2. 5 per 100

3. 13:100

4. $\frac{39}{100}$

5. 12.5 per 100

6. 51 out of 100

Write each fraction as a percent.

7. $\frac{7}{10}$

8. $\frac{6}{50}$

9. $\frac{13}{20}$

10. $\frac{30}{50}$

11. $\frac{7}{20}$

12. $\frac{12}{20}$

13. $\frac{23}{25}$

14. $\frac{3}{10}$

15. $\frac{17}{50}$

Write each percent as a fraction in simplest form.

16. 15%

17. 85%

18. 1%

19. 70%

20. 25%

21. 19%

22. 33%

23. 22%

24. 95%

NAME ______________________ DATE ____________ PERIOD _____

4-7 Skills Practice

Percents and Decimals

Write each percent as a decimal.

1. 5%

2. 20%

3. 21%

4. 83%

5. 7%

6. 56%

7. 16%

8. 45%

9. 27.3%

10. 14.9%

11. 91.5%

12. 29.3%

13. 14.4%

14. 80%

15. 7.5%

16. $10\frac{1}{2}\%$

Write each decimal as a percent.

17. 0.06

18. 0.13

19. 0.5

20. 0.74

21. 0.14

22. 0.92

23. 0.54

24. 0.66

25. 0.192

26. 0.295

27. 0.911

28. 0.247

29. 0.4165

30. 0.2199

31. 0.7601

32. 0.4833

NAME ______________________ DATE ____________ PERIOD _____

4-8 Skills Practice

Least Common Multiple

Find the LCM of each set of numbers.

1. 2, 8

2. 6, 10

3. 10, 11

4. 10, 12

5. 9, 18

6. 4, 22

7. 12, 30

8. 4, 13

9. 25, 30

10. 250, 30

11. 200, 18

12. 70, 90

13. 18, 54

14. 30, 65

15. 180, 252

16. 20, 55

17. 21, 60

18. 3, 5, 10

19. 3, 4, 13

20. 4, 10, 12

21. 6, 15, 20

22. 48, 16, 3

23. 66, 55, 44

24. 29, 58, 4

NAME ______________________ DATE ____________ PERIOD _____

4-9 Skills Practice

Comparing and Ordering Rational Numbers

Find the LCD of each pair of fractions.

1. $\frac{4}{7}, \frac{3}{5}$

2. $\frac{5}{12}, \frac{7}{24}$

3. $\frac{6}{28}, \frac{3}{7}$

4. $\frac{7}{15}, \frac{1}{4}$

5. $\frac{7}{11}, \frac{3}{5}$

6. $\frac{5}{17}, \frac{7}{8}$

7. $\frac{5}{12}, \frac{7}{10}$

8. $\frac{15}{16}, \frac{1}{4}$

9. $\frac{5}{8}, \frac{3}{5}$

Replace each ● with <, >, or = to make a true sentence.

10. $\frac{3}{10}$ ● $\frac{2}{9}$

11. $\frac{3}{7}$ ● $\frac{5}{11}$

12. $\frac{9}{12}$ ● $\frac{3}{4}$

13. $\frac{12}{13}$ ● $\frac{14}{15}$

14. $\frac{4}{5}$ ● $\frac{5}{4}$

15. $\frac{17}{30}$ ● $\frac{13}{20}$

16. $\frac{35}{60}$ ● $\frac{49}{84}$

17. $3\frac{4}{11}$ ● $3\frac{7}{20}$

18. $1\frac{2}{3}$ ● $\frac{9}{7}$

Order each set of ratios from least to greatest.

19. 0.48, 0.46, $\frac{9}{20}$

20. 0.99, 0.89, $\frac{7}{8}$

21. $\frac{1}{4}$, 0.2, 0.4

Determine whether each number is rational. Write *yes* or *no*.

22. 2.323323332...

23. $\frac{7}{19}$

24. $4.\overline{3}$

NAME ______________________ DATE ____________ PERIOD ______

5-1

Skills Practice

Estimating with Fractions

Estimate.

1. $\frac{4}{5} + \frac{2}{11}$

2. $\frac{4}{7} + \frac{1}{5}$

3. $\frac{7}{9} - \frac{1}{5}$

4. $\frac{9}{10} - \frac{1}{23}$

5. $\frac{3}{5} + \frac{9}{11}$

6. $\frac{4}{5} - \frac{4}{9}$

7. $5\frac{1}{7} + 7\frac{9}{11}$

8. $3\frac{10}{11} - 2\frac{1}{6}$

9. $5\frac{1}{4} - \frac{1}{7}$

10. $8\frac{3}{7} - 2\frac{1}{2}$

11. $2\frac{1}{8} + 6\frac{9}{10}$

12. $10\frac{1}{8} - 3\frac{1}{4}$

13. $\frac{4}{5} \times \frac{8}{9}$

14. $\frac{6}{7} \div \frac{10}{11}$

15. $3\frac{6}{7} \times 2\frac{1}{10}$

16. $16\frac{1}{3} \div 3\frac{8}{9}$

17. $31\frac{3}{4} \div 2\frac{1}{8}$

18. $3\frac{4}{5} \cdot 1\frac{1}{4}$

19. $12 \div 2\frac{6}{7}$

20. $44\frac{1}{5} \div 3\frac{7}{8}$

21. $10\frac{1}{7} \cdot 4\frac{1}{3}$

22. $5\frac{1}{8} \cdot 6\frac{9}{11}$

23. $\frac{3}{12} \div 4\frac{4}{5}$

24. $2\frac{1}{2} \div 3\frac{1}{3}$

25. Estimate $36\frac{1}{4}$ divided by 6.

26. Estimate the sum of $7\frac{9}{10}$, $2\frac{1}{5}$, and $3\frac{2}{3}$.

NAME ______________________ DATE ____________ PERIOD _____

5-2 Skills Practice

Adding and Subtracting Fractions

Add or subtract. Write in simplest form.

1. $\frac{3}{8} + \frac{3}{8}$

2. $\frac{7}{10} - \frac{5}{10}$

3. $\frac{9}{10} + \frac{3}{10}$

4. $\frac{4}{7} - \frac{2}{7}$

5. $\frac{2}{3} + \frac{2}{3}$

6. $\frac{5}{9} - \frac{2}{9}$

7. $\frac{8}{15} - \frac{1}{5}$

8. $\frac{5}{6} + \frac{5}{12}$

9. $\frac{3}{5} - \frac{3}{10}$

10. $\frac{7}{16} + \frac{3}{8}$

11. $\frac{19}{20} - \frac{3}{10}$

12. $\frac{5}{9} + \frac{7}{9}$

13. $\frac{4}{9} - \frac{1}{12}$

14. $\frac{2}{3} + \frac{3}{7}$

15. $\frac{3}{4} + \frac{1}{7}$

16. $\frac{7}{8} - \frac{2}{3}$

17. $\frac{8}{9} - \frac{5}{6}$

18. $\frac{5}{12} - \frac{3}{10}$

19. $\frac{7}{9} + \frac{2}{3}$

20. $\frac{3}{5} + \frac{4}{11}$

21. $\frac{11}{12} - \frac{1}{4}$

ALGEBRA Evaluate each expression if $a = \frac{5}{6}$ and $b = \frac{3}{8}$.

22. $a + b$

23. $a - b$

24. $\frac{9}{10} - a$

NAME ______________________ DATE __________ PERIOD _____

5-3 Skills Practice

Adding and Subtracting Mixed Numbers

Lesson 5-3

Add or subtract. Write in simplest form.

1. $3\frac{2}{7} + 2\frac{1}{7}$

2. $7\frac{1}{3} + 7\frac{1}{3}$

3. $9\frac{3}{5} - 2\frac{1}{5}$

4. $7\frac{3}{4} - 5\frac{1}{4}$

5. $3\frac{1}{4} + 5\frac{1}{4}$

6. $6\frac{3}{4} - 5\frac{3}{4}$

7. $12\frac{1}{8} + 9\frac{3}{8}$

8. $5\frac{2}{3} - 2\frac{1}{3}$

9. $14\frac{3}{5} - 9\frac{2}{5}$

10. $5\frac{1}{2} + 3\frac{1}{4}$

11. $2\frac{1}{3} + 6\frac{1}{6}$

12. $6\frac{1}{3} - 6\frac{1}{4}$

13. $7\frac{5}{6} - 2\frac{2}{3}$

14. $6\frac{7}{10} + 5\frac{1}{4}$

15. $12\frac{3}{8} - 9\frac{1}{3}$

16. $12\frac{13}{15} + 4\frac{1}{9}$

17. $15\frac{2}{3} - 7\frac{1}{5}$

18. $4\frac{7}{12} - 2\frac{3}{16}$

19. $8\frac{3}{4} + 3\frac{2}{5}$

20. $12\frac{1}{3} - 8\frac{5}{9}$

21. $8 - 3\frac{2}{5}$

22. $7\frac{7}{9} + 6\frac{7}{8}$

23. $7\frac{7}{9} - 6\frac{7}{8}$

24. $10\frac{3}{8} + 7\frac{11}{12}$

NAME ______________________ DATE ____________ PERIOD _____

5-4 Skills Practice

Problem-Solving Investigation: Eliminate Possibilities

Eliminate possibilities to solve.

1. **CELL PHONES** Justin has a cell phone plan that costs $25 per month plus $0.05 per minute for each minute he talks. How much will it cost Justin for the month of March if he talked for 120 minutes?
   A $25 **C** $31
   B $25.05 **D** $22

2. **PUPPIES** A puppy gains weight at a rate of 1.5 pounds every two weeks for the first 6 months after it is born. If Sam's puppy weighs 6 pounds right now, how much will it weigh six weeks from now?
   F 6 pounds **H** 15 pounds
   G 10.5 pounds **J** 13 pounds

3. **RECIPES** Ryan needs $3\frac{1}{4}$ cups of flour for the cookie recipe she is baking. She has already added $2\frac{1}{2}$ cups of flour. How much more flour does she need to add?
   A $2\frac{1}{2}$ cups **C** $3\frac{1}{4}$ cups
   B 1 cup **D** $\frac{3}{4}$ cup

4. **HOMEWORK** Jordan does 30 minutes of homework per day. How many hours of homework does she do in one week?
   F $3\frac{1}{2}$ hours **H** $\frac{1}{2}$ hour
   G 2 hours **J** 3 hours

5. **FENCING** A rectangular garden is surrounded by a fence. The length of the garden is $3\frac{1}{2}$ feet and the width of the garden is $1\frac{1}{2}$ feet. How much fencing must be used to surround the garden?
   A 5 feet **C** 10 feet
   B $3\frac{1}{2}$ feet **D** $1\frac{1}{2}$ feet

6. **PARTIES** The 6th grade class is renting the pool at the recreation center for a holiday party. It costs $75 plus $1.50 per person to rent the pool. How much will it cost to rent the pool if 120 people plan on attending ?
   F $76.50 **H** $255
   G $195 **J** $120

NAME ______________________ DATE ____________ PERIOD _____

5-5 Skills Practice

Multiplying Fractions and Mixed Numbers

Multiply. Write in simplest form.

1. $\frac{1}{2} \times \frac{4}{5}$

2. $\frac{1}{9} \times \frac{3}{5}$

3. $\frac{15}{24} \times \frac{3}{20}$

4. $\frac{1}{7} \times \frac{1}{5}$

5. $\frac{5}{7} \times \frac{14}{15}$

6. $\frac{9}{10} \times \frac{5}{9}$

7. $\frac{4}{11} \times \frac{3}{8}$

8. $\frac{2}{3} \times \frac{7}{9}$

9. $\frac{9}{13} \times \frac{26}{27}$

10. $\frac{4}{9} \times 5$

11. $7 \times \frac{2}{7}$

12. $2\frac{4}{5} \times \frac{1}{3}$

13. $4\frac{1}{2} \times \frac{1}{3}$

14. $5\frac{3}{4} \times 12$

15. $14 \times 2\frac{3}{7}$

16. $2\frac{3}{5} \times 1\frac{3}{7}$

17. $1\frac{4}{9} \times 2\frac{4}{7}$

18. $5\frac{5}{6} \times 6\frac{3}{8}$

19. $10\frac{7}{9} \times 4\frac{1}{4}$

20. $9\frac{7}{9} \times 7\frac{3}{4}$

21. $3\frac{3}{4} \times 2\frac{4}{7}$

Lesson 5-5

NAME ______________________ DATE ____________ PERIOD _____

5-6 Skills Practice

Algebra: Solving Equations

Find the multiplicative inverse of each number.

1. $\frac{3}{7}$

2. $-\frac{4}{11}$

3. $\frac{7}{2}$

4. $\frac{9}{5}$

5. –5

6. $6\frac{1}{3}$

7. $4\frac{1}{9}$

8. $17\frac{1}{2}$

9. $-15\frac{2}{3}$

Solve each equation. Check your solution.

10. $\frac{x}{10} = 3$

11. $7 = \frac{m}{4}$

12. $\frac{a}{5} = -11$

13. $-8 = \frac{t}{5}$

14. $\frac{3}{4}c = 12$

15. $-7 = -\frac{w}{6}$

16. $\frac{3}{5}y = 6$

17. $15 = \frac{3}{7}b$

18. $\frac{6}{7}c = 18$

19. $\frac{7}{3}x = \frac{2}{3}$

20. $\frac{11}{12} = \frac{3}{4}h$

21. $\frac{9}{14}y = \frac{3}{7}$

22. $\frac{m}{2.6} = -5$

23. $0.6 = \frac{n}{5}$

24. $\frac{r}{2.6} = -1.3$

NAME ______________________ DATE ____________ PERIOD _____

5-7 Skills Practice

Dividing Fractions and Mixed Numbers

Divide. Write in simplest form.

1. $\frac{1}{6} \div \frac{1}{5}$

2. $5 \div \frac{3}{5}$

3. $\frac{6}{7} \div \frac{1}{7}$

4. $\frac{3}{4} \div \frac{1}{2}$

5. $8 \div \frac{1}{3}$

6. $\frac{1}{5} \div \frac{1}{4}$

7. $7 \div \frac{3}{7}$

8. $\frac{4}{7} \div \frac{8}{9}$

9. $8\frac{1}{3} \div 5$

10. $\frac{9}{7} \div \frac{3}{14}$

11. $\frac{12}{5} \div \frac{3}{10}$

12. $5 \div 3\frac{3}{4}$

13. $6\frac{4}{5} \div 17$

14. $7\frac{1}{3} \div 4$

15. $\frac{3}{4} \div 5\frac{1}{2}$

16. $\frac{2}{7} \div 1\frac{13}{14}$

17. $\frac{3}{8} \div 6\frac{1}{4}$

18. $7\frac{1}{2} \div 2\frac{5}{6}$

19. $3\frac{4}{9} \div 2\frac{1}{3}$

20. $2\frac{2}{3} \div 1\frac{1}{6}$

21. $4\frac{3}{4} \div 2\frac{1}{2}$

Lesson 5-7

NAME ________________________ DATE ____________ PERIOD _____

6-1 Skills Practice

Ratios

Write each ratio as a fraction in simplest form.

1. 14 to 6

2. 18:3

3. 4:22

4. 7:21

5. 18:12

6. 20 to 9

7. 25 to 20

8. 4:10

9. 18:21

10. 84 to 16

11. 33 ounces to 11 ounces

12. 45 minutes:25 minutes

13. 77 cups:49 cups

14. 15 pounds to 39 pounds

15. 40 seconds to 60 seconds

16. 140 centimeters to 300 centimeters

17. 9 weeks: 15 weeks

18. 3 yards to 33 yards

Determine whether the ratios are equivalent. Explain.

19. $\frac{3}{16}$ and $\frac{9}{48}$

20. $\frac{7}{10}$ and $\frac{8}{11}$

21. 18 in.:3 ft and 12 in.:2 ft

22. 6 mos:2 yr and 8 mos:3 yr

NAME ______________________ DATE ____________ PERIOD ____

6-2 Skills Practice

Rates

Find each unit rate. Round to the nearest hundredth if necessary.

1. $112 in 8 hours

2. 150 miles in 6 gallons

3. 49 points in 7 games

4. 105 students in 3 classes

5. 120 problems in 5 hours

6. 3 accidents in 12 months

7. 6 eggs in 7 days

8. 8 batteries in 3 months

9. 122 patients in 4 weeks

10. 51 gallons in 14 minutes

11. $8.43 for 3 pounds

12. 357 miles in 6.3 hours

13. 25 letters in 4 days

14. $99 for 12 CDs

15. 5 breaks in 8 hours

16. 3 trips in 14 months

17. 2 pay raises in 3 years

18. 7 errors in 60 minutes

19. 15 pounds in 6 weeks

20. 8 commercials in 15 minutes

21. 8 glasses every 24 hours

22. 13 feet in 5 steps

Choose the better unit price.

23. $4.99 for 6 cans or $7.99 for 10 cans

24. $21.50 for 4 pounds of lunch meat or $15.10 for 3 pounds of lunch meat

NAME ______________________ DATE ____________ PERIOD _____

6-3 Skills Practice

Rate of Change and Slope

Find the rate of change for each table.

1.

Time spent Mowing (in hours)	Money Earned (in dollars)
1	10
3	30
5	50
7	70

2.

Time (in hours)	Temperature (in degrees)
9:00	60
10:00	62
11:00	64
12:00	66

3.

Number of Students	Number of Magazines Sold
10	100
15	150
20	200
25	250

4.

Number of Trees	Number of Apples
5	100
10	200
15	300
20	400

5.

Number of Volunteers	Number of Hours Logged
5	10
10	20
15	30
20	40

6.

Gas Left in Tank	Miles Driven
12	0
10	50
8	100
6	150

Find the rate of change for each graph.

7.

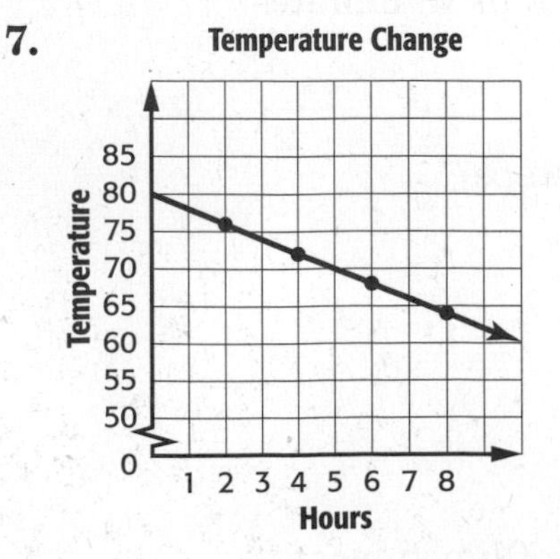

8.

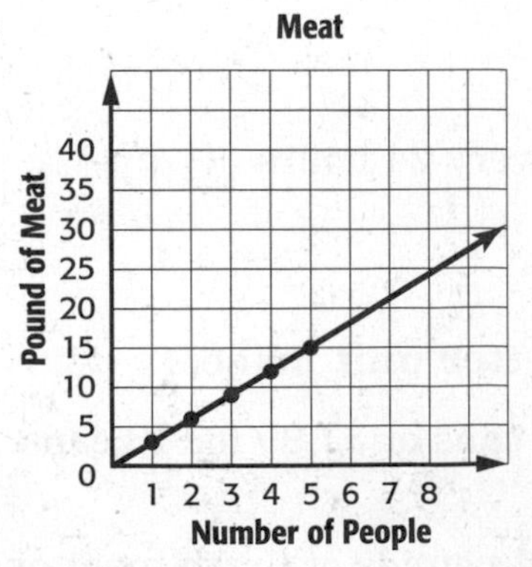

NAME ______________ DATE ______ PERIOD ____

6-4

Skills Practice

Measurement: Changing Customary Units

Complete.

1. 3 lb = ________ oz

2. 16 qt = ________ gal

3. 24 in. = ________ ft

4. 12 ft = ________ yd

5. 3 mi = ________ ft

6. 12,000 lb = ________ T

7. 64 oz = ________ lb

8. 6 pt = ________ qt

9. 3 pt = ________ c

10. $5\frac{1}{2}$ ft = ________ in.

11. 22 yd = ________ ft

12. $\frac{1}{4}$ mi = ________ ft

13. 15 T = ________ lb

14. 7 lb = ________ oz

15. $8\frac{1}{2}$ qt = ________ pt

16. 5 gal = ________ qt

17. 8 c = ________ pt

18. 16 in = ________ ft

19. 24 fl oz = ________ c

20. 60 ft = ________ yd

21. 6,600 ft = ________ mi

22. 7.5 T = ________ lb

23. 88 oz = ________ lb

24. 70 qt = ________ gal

Lesson 6-4

NAME ______________ DATE ______________ PERIOD ______

6-5 Skills Practice

Measurement: Changing Metric Units

Complete.

1. 660 m = ____ km

2. 5.7 m = ____ cm

3. 543 mL = ____ L

4. 23.7 g = ____ mg

5. 529 mg = ____ g

6. 2,640 mL = ____ L

7. 4.32 kL = ____ L

8. 75.4 mg = ____ g

9. 8,300 mg = ____ g

10. 7.3 m = ____ cm

11. 250.3 kL = ____ L

12. 799 g = ____ kg

13. 8.5 cm = ____ mm

14. 450 kg = ____ g

15. 7.3 L = ____ mL

16. 6,140 L = ____ kL

17. 3,500 m = ____ km

18. 89 km = ____ m

19. 26.8 mm = ____ cm

20. 750 m = ____ km

21. 4.8 m = ____ cm

22. 95 g = ____ mg

23. 389 mm = ____ m

24. 56 L = ____ kL

25. 0.32 mm = ____ cm

26. 39.1 g = ____ kg

NAME ______________________ DATE ____________ PERIOD _____

6-6 Skills Practice

Algebra: Solving Proportions

Determine if the quantities in each pair of ratios are proportional.

1. $\frac{9}{5} = \frac{27}{15}$

2. $\frac{16}{10} = \frac{24}{15}$

3. $\frac{6}{18} = \frac{9}{25}$

4. $\frac{42}{63} = \frac{28}{42}$

5. $\frac{11}{8} = \frac{13}{10}$

6. $\frac{22}{33} = \frac{12}{18}$

7. $\frac{14}{17} = \frac{29}{35}$

8. $\frac{36}{22} = \frac{30}{19}$

9. $\frac{32}{48} = \frac{10}{15}$

10. $\frac{320 \text{ mi}}{6 \text{ h}} = \frac{420 \text{ mi}}{8 \text{ h}}$

11. $\frac{\$4.96}{8 \text{ oz}} = \frac{\$3.72}{6 \text{ oz}}$

12. $\frac{25 \text{ mg}}{1.5 \text{ c}} = \frac{100 \text{ mg}}{6 \text{ c}}$

Solve each proportion.

13. $\frac{24}{13} = \frac{a}{26}$

14. $\frac{18}{x} = \frac{3}{36}$

15. $\frac{3}{u} = \frac{5}{15}$

16. $\frac{650}{6.5} = \frac{z}{5}$

17. $\frac{2.8}{4} = \frac{7}{q}$

18. $\frac{c}{17} = \frac{0.01}{8.5}$

19. $\frac{0.1}{8.2} = \frac{1.8}{b}$

20. $\frac{300}{24} = \frac{18}{j}$

21. $\frac{4.2}{t} = \frac{8}{5}$

22. $\frac{120}{75} = \frac{8}{m}$

Lesson 6-6

6-7 Skills Practice

Problem-Solving Investigation: Draw a Diagram

Draw a diagram to solve.

1. **HOMEWORK** Shantel is studying for her history test. After 20 minutes, she is $\frac{1}{4}$ of the way done. How much longer will she study?

2. **RECIPES** Damon is making muffins. He has added $\frac{3}{4}$ of the ingredients. If he has added 6 ingredients, how many more does he have to add to be finished?

3. **TRAVEL** The Smithsons are going to Dallas, TX on vacation. They have traveled $\frac{1}{3}$ of the total distance. If they have traveled 126 miles, how far is it from their house to Dallas?

4. **PHYSICS** A ball is dropped from 256 feet above the ground. It bounces up $\frac{1}{4}$ as high as it fell. This is true for each successive bounce. What height will the ball reach on the third bounce?

5. **SCHOOL** Mrs. Wright says that $\frac{2}{3}$ of her class has arrived for the day. If 10 students have arrived, how many students are in her class?

6. **TRAVEL** Jeremy walked $\frac{1}{4}$ of the way to school, ran $\frac{1}{4}$ of the way to school, then rode with his best friend the rest of the way. If he walked 1.5 miles, how far did he ride with his friend?

NAME ______________________ DATE ____________ PERIOD _____

6-8 Skills Practice

Scale Drawings

Lesson 6-8

ARCHITECTURE The scale on a set of architectural drawings for a house is $\frac{1}{2}$ inch = $1\frac{1}{2}$ feet. Find the length of each part of the house.

	Room	Drawing Length	Actual Length
1.	Living Room	5 inches	
2.	Dining Room	4 inches	
3.	Kitchen	$5\frac{1}{2}$ inches	
4.	Laundry Room	$3\frac{1}{4}$ inches	
5.	Basement	10 inches	
6.	Garage	$8\frac{1}{3}$ inches	

ARCHITECTURE As part of a city building refurbishment project, architects have constructed a scale model of several city buildings to present to the city commission for approval. The scale of the model is 1 inch = 9 feet.

7. The courthouse is the tallest building in the city. If it is $7\frac{1}{2}$ inches tall in the model, how tall is the actual building?

8. The city commission would like to install new flagpoles that are each 45 feet tall. How tall are the flagpoles in the model?

9. In the model, two of the flagpoles are 4 inches apart. How far apart will they be when they are installed?

10. The model includes a new park in the center of the city. If the dimensions of the park in the model are 9 inches by 17 inches, what are the actual dimensions of the park?

11. Find the scale factor.

NAME ______________________ DATE ____________ PERIOD ______

6-9 Skills Practice

Fractions, Decimals, and Percents

Write each percent as a fraction in simplest form.

1. 18%

2. 67.5%

3. 21.25%

4. 87.5%

5. $31\frac{1}{4}\%$

6. 17.5%

7. $18\frac{3}{4}\%$

8. $68\frac{3}{4}\%$

9. 7.5%

10. 130%

11. 0.5%

12. 0.02%

Write each fraction as a percent. Round to the nearest hundredth if necessary.

13. $\frac{3}{5}$

14. $\frac{3}{8}$

15. $\frac{2}{18}$

16. $\frac{3}{16}$

17. $\frac{7}{9}$

18. $\frac{21}{50}$

19. $\frac{1}{3}$

20. $\frac{40}{42}$

21. $\frac{7}{16}$

22. $\frac{17}{10}$

23. $\frac{1}{500}$

24. $\frac{26}{25}$

NAME ______________________ DATE ____________ PERIOD _____

7-1 Skills Practice

Percent of a Number

Find each number.

1. Find 80% of 80.

2. What is 95% of 600?

3. 35% of 20 is what number?

4. Find 60% of $150.

5. What is 75% of 240?

6. 380% of 30 is what number?

7. Find 40% of 80.

8. What is 30% of $320?

9. 12% of 150 is what number?

10. Find 58% of 200.

11. What is 18% of $450?

12. What is 70% of 1,760?

13. Find 92% of 120.

14. 45% of 156 is what number?

15. What is 12% of 12?

16. Find 60% of 264.

17. 37.5% of 16 is what number?

18. What is 82.5% of 400?

19. What is 0.25% of 900?

20. Find 1.5% of 220.

NAME ______________________ DATE ____________ PERIOD _____

7-2 Skills Practice

The Percent Proportion

Find each number. Round to the nearest tenth if necessary.

1. 50 is 20% of what number?

2. What percent of 20 is 4?

3. What number is 70% of 250?

4. 10 is 5% of what number?

5. What number is 45% of 180?

6. 40% of what number is 82?

7. What percent of 90 is 36?

8. 60 is 25% of what number?

9. What number is 32% of 1,000?

10. What percent of 125 is 5?

11. 73 is 20% of what number?

12. 57% of 109 is what number?

13. What percent of 185 is 35?

14. 25 is what percent of 365?

15. 85% of 190 is what number?

16. 12.5 is 25% of what number?

17. What percent of 128 is 24?

18. 5.25% of 170 is what number?

19. What is 82% of 230?

20. What percent of 49 is 7?

7-3 Skills Practice

Percent and Estimation

Estimate by using fractions.

1. 51% of 128

2. 76% of 200

3. 32.9% of 90

4. 23% of 8

5. 19% of 45

6. 81% of 16

Estimate by using 10%.

7. 12% of 98

8. 89% of 300

9. 31% of 80

10. 28% of 49

11. 62% of 13

12. 77% of 28

Estimate.

13. 308% of 500

14. 0.5% of 87

15. 153% of 20

16. 0.6% of 41

17. 231% of 54

18. 0.9% of 116

19. 0.26% of 36

20. 425% of 119

NAME ______________________ DATE ____________ PERIOD _____

7-4 Skills Practice

Algebra: The Percent Equation

Write an equation for each problem. Then solve. Round to the nearest tenth if necessary.

1. 25% of 176 is what number?

2. What is 90% of 20?

3. 24 is what percent of 30?

4. 80% of what number is 94?

5. What is 60% of 45?

6. 9 is what percent of 30?

7. What percent of 125 is 25?

8. What is 120% of 20?

9. 2% of what number is 5?

10. 15% of 290 is what number?

11. 16 is what percent of 4,000?

12. What is 140% of 60?

13. 344.8 is what percent of 862?

14. 6% of what number is 21?

15. What number is 60% of 605?

16. 32% of 250 is what number?

17. Find 30% of 70.

18. What is 80% of 65?

NAME ______________________ DATE ____________ PERIOD _____

7-5 Skills Practice

Problem-Solving Investigation: Determine Reasonable Answers

Determine reasonable answers for each.

1. **MONEY** Gillian and Roger have lunch at a restaurant and Gillian needs to determine how much tip to leave based on their bill. If their bill was \$21.87 and Gillian wants to leave a 15% tip, what is a reasonable estimate for how much she should leave?

2. **SPORTS** Of the 82,000 fans that attended a bowl game between Ohio State and Notre Dame, 60% were Ohio State fans. About how many fans at the game were for Notre Dame?

3. **ICE CREAM** A survey of 1,950 people found that 39% preferred chocolate ice cream to vanilla. About how many people preferred chocolate ice cream according to the survey?

4. **EARTH** The surface area of Earth is approximately 70% water. If the surface area is about 510,000,000 square kilometers, about how many square kilometers are water?

5. **COLLEGE** Of 7,450 first-year college students interviewed, 72% had changed their major area of study since the beginning of the academic year. About how many students had kept the same major?

6. **MONEY** While shopping, Hilary spent \$149. If the amount she spent was 20% of her savings, how much savings did she have before she shopped?

Lesson 7-5

7-6 Skills Practice

Percent of Change

Find each percent of change. Round to the nearest whole percent if necessary. State whether the percent of change is an *increase* or *decrease*.

1. original: 35
new: 70

2. original: 8
new: 12

3. original: 45
new: 30

4. original: $350
new: $400

5. original: $75
new: $60

6. original: 250
new: 100

7. original: $30
new: $110

8. original: 35
new: 28

9. original: $12.50
new: $15

10. original: 80
new: 52

11. original: 45
new: 63

12. original: 120
new: 132

13. original: $210
new: $105

14. original: 84
new: 111

15. original: $84
new: $100

16. original: 6.8
new: 8.2

17. original: 1.5
new: 2.5

18. original: 91
new: 77

19. original: $465.50
new: $350

20. original: $87.05
new: $100

21. original: 144
new: 108

22. original: 20.8
new: 12.2

23. original: $75
new: $15

24. original: 8.6
new: 7

NAME ____________________ DATE __________ PERIOD _____

7-7 Skills Practice

Sales Tax and Discount

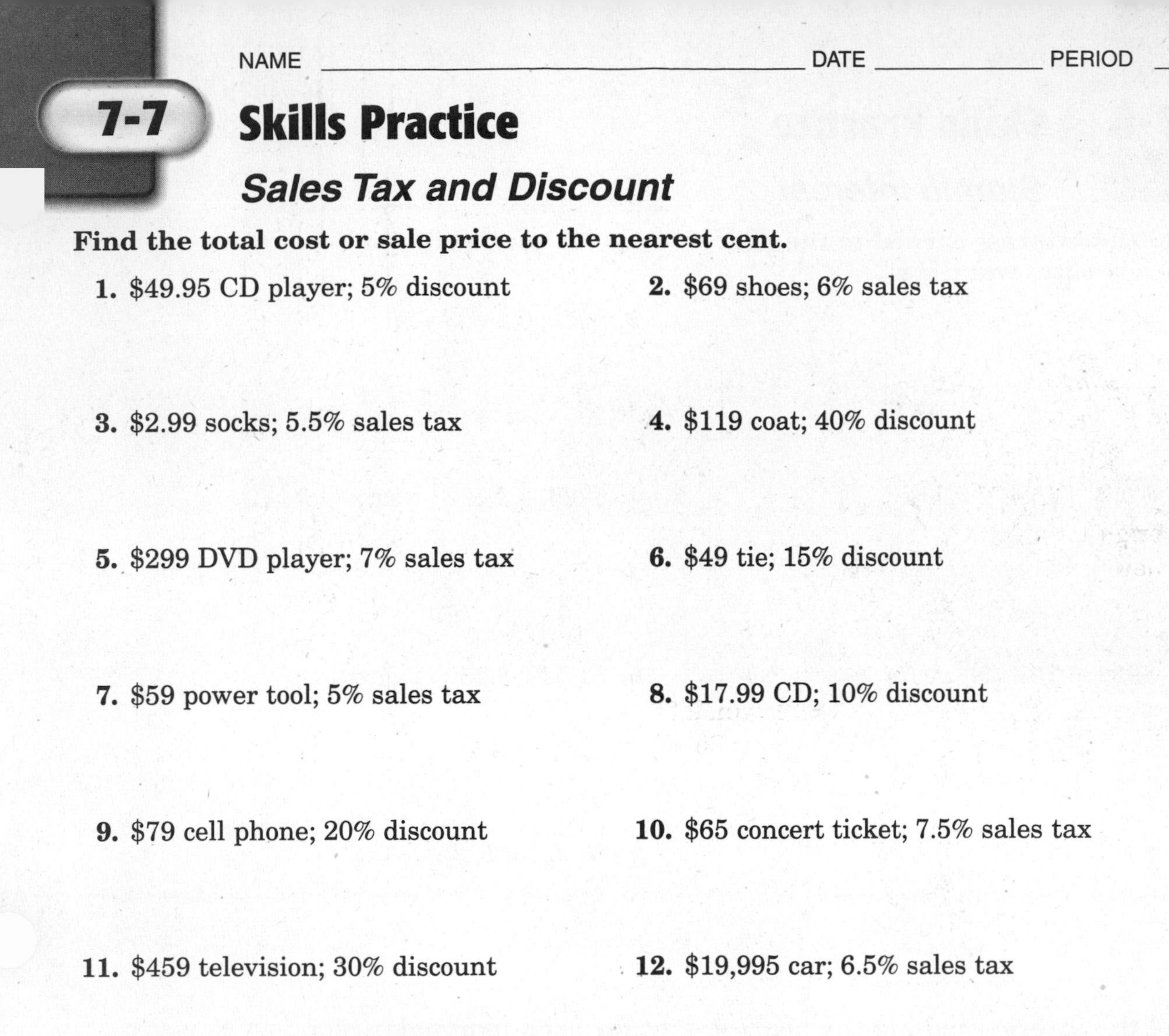

Find the total cost or sale price to the nearest cent.

1. $49.95 CD player; 5% discount

2. $69 shoes; 6% sales tax

3. $2.99 socks; 5.5% sales tax

4. $119 coat; 40% discount

5. $299 DVD player; 7% sales tax

6. $49 tie; 15% discount

7. $59 power tool; 5% sales tax

8. $17.99 CD; 10% discount

9. $79 cell phone; 20% discount

10. $65 concert ticket; 7.5% sales tax

11. $459 television; 30% discount

12. $19,995 car; 6.5% sales tax

Find the original price to the nearest cent.

13. boots: discount, 30%
sale price, $62.50

14. video game: discount, 15%
sale price, $12.64

15. drum set: discount, 10%
sale price, $1,099

16. gloves: discount, 30%
sale price, $16.40

17. sweater: discount, 30%
sale price, $34

18. sunglasses: discount, 20%
sale price, $62.95

19. dinner for two: discount, 5%
sale price, $70

20. bicycle: discount, 25%
sale price, $147.85

Lesson 7-7

NAME ______________________ DATE ____________ PERIOD _____

7-8 Skills Practice

Simple Interest

Find the interest earned to the nearest cent for each principal, interest rate, and time.

1. \$500, 4%, 2 years

2. \$350, 6.2%, 3 years

3. \$740, 3.25%, 2 years

4. \$725, 4.3%, $2\frac{1}{2}$ years

5. \$955, 6.75%, $3\frac{1}{4}$ years

6. \$1,540, 8.25%, 2 years

7. \$3,500, 4.2%, $1\frac{3}{4}$ years

8. \$568, 16%, 8 months

Find the interest paid to the nearest cent for each loan balance, interest rate, and time.

9. \$800, 9%, 4 years

10. \$280, 5.5%, 4 years

11. \$1,150, 7.6%, 5 years

12. \$266, 5.2%, 3 years

13. \$450, 22%, 1 year

14. \$2,180, 7.7%, $2\frac{1}{2}$ years

15. \$2,650, 3.65%, $4\frac{1}{2}$ years

16. \$1,245, 5.4%, 6 months

NAME ______________ DATE ________ PERIOD _____

8-1 Skills Practice

Line Plots

For Exercises 1–3, use the data at the right that shows the number of fish each person caught on a fishing trip.

Number of Fish				
3	1	0	1	0
1	2	3	1	4
2	1	2	3	0
1	2	3	2	7

1. Make a line plot of the data.

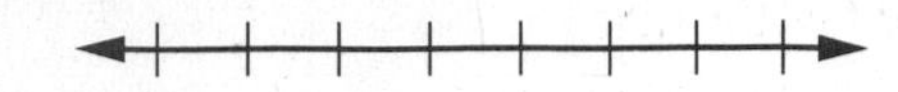

2. What is the range of the data?

3. Identify any clusters, gaps, or outliers and analyze the data by describing what these values represent.

Make a line plot for each set of data. Identify any clusters, gaps, or outliers.

4.

Test Scores			
83	84	92	91
82	81	80	94
85	95	96	84
94	98	93	90

5.

Rainfall (in.)			
3	2	4	3
1	8	7	3
2	9	4	0

For Exercises 6–8, use the line plot at the right.

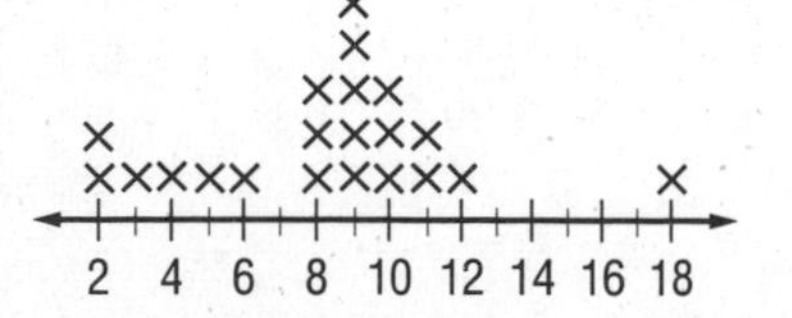

6. What is the range of the data?

7. What number occurred most often?

8. Identify any clusters, gaps, or outliers.

Lesson 8-1

NAME ______________________ DATE ____________ PERIOD _____

8-2 Skills Practice

Measures of Central Tendency and Range

Find the mean, median, and mode for each set of data. Round to the nearest tenth if necessary.

1. 5, 9, 6, 6, 11, 8, 4

2. 1, 3, 5, 2, 4, 8, 4, 7, 2

3. 1, 9, 4, 7, 5, 3, 16, 11

4. 3, 4, 4, 4, 4, 3, 6

5. 3, 7, 2, 5, 5, 6, 5, 10, 11, 5

6. 19, 17, 24, 11, 19, 25, 15, 15, 19, 16, 16

7. 5, 8, 9, 9, 12, 6, 4

8. 3, 4, 9, 7, 6, 6, 2

9.

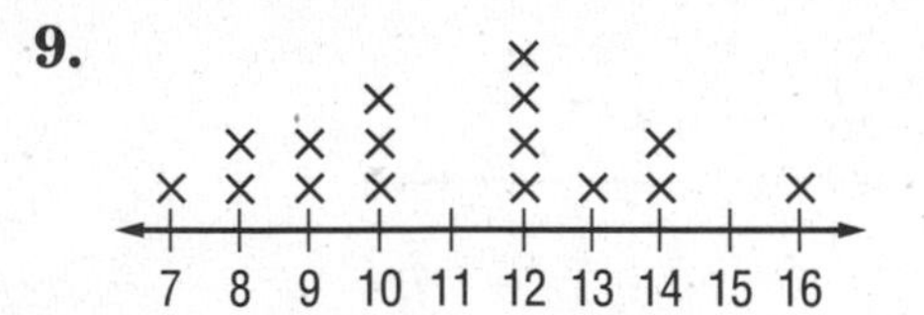

10.

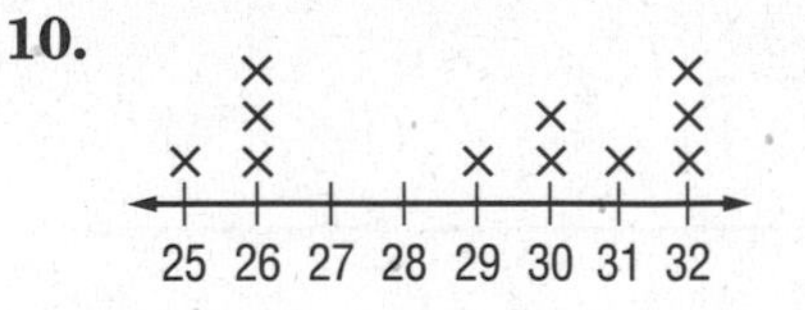

11.

× × × × × ×

14 15 16 17 18 19 20

12.

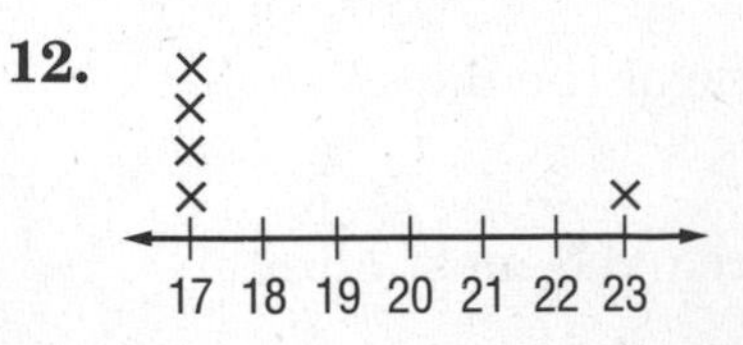

NAME ______________________ DATE ____________ PERIOD _____

8-3 Skills Practice

Stem-and-Leaf Plots

Make a stem-and-leaf plot for each set of data.

1. 23, 36, 25, 13, 24, 25, 32, 33, 17, 26, 24

2. 3, 4, 6, 17, 12, 5, 17, 4, 26, 17, 18, 21, 16, 15, 20

3. 26, 27, 23, 23, 24, 26, 31, 45, 33, 32, 41 40, 21, 20

4. 347, 334, 346, 330, 348, 347, 359, 344, 357

HOT DOGS **For Exercises 5–7, use the stem-and-leaf plot at the right that shows the number of hot dogs eaten during a contest.**

Stem	Leaf
0	8 8 9
1	1 2 2 4 7 7 7
2	1 1 2

$2|1 = 21$

5. How many hot dogs are represented on the stem-and-leaf plot?

6. What is the range of the number of hot dogs eaten?

7. Find the median and mode of the data.

Determine the mean, median, and mode of the data shown in each stem-and-leaf plot.

8.

Stem	Leaf
0	1 2 2 3
1	3 4 5 5
2	0 0 0 1 3

$2|0 = 20$

9.

Stem	Leaf
2	0 0 0 2 3 5 7
3	1 2
4	0

$4|0 = 40$

10.

Stem	Leaf
22	1 1 2 7
23	3 3 9
24	0 6 8

$24|0 = 240$

11.

Stem	Leaf
0	1 3 3 4 7
1	2 2 2 4 5 6
2	0 0 0 1

$2|0 = 20$

8-4 Skills Practice

Bar Graphs and Histograms

ZOOS For Exercises 1 and 2, use the table. It shows the number of species at several zoological parks.

Zoo	Species
Los Angeles	350
Lincoln Park	290
Cincinnati	700
Bronx	530
Oklahoma City	600

1. Make a bar graph of the data.

Animal Species in Zoos

2. Which zoological park has the most species?

ZOOS For Exercises 3 and 4, use the table at the right. It shows the number of species at 37 major U.S. public zoological parks.

Number of Species				
200	700	290	600	681
300	643	350	794	400
360	600	134	200	800
305	384	500	330	250
530	715	303	200	475
465	340	347	300	708
184	800	375	350	450
337	221			

3. Make a histogram of the data. Use intervals of 101–200, 201–300, 301–400, 401–500, 501–600, 601–700, and 701–800 for the horizontal axis.

Animal Species in Zoos

4. Which interval has the largest frequency?

HEALTH For Exercises 5 and 6, use the graph at the right.

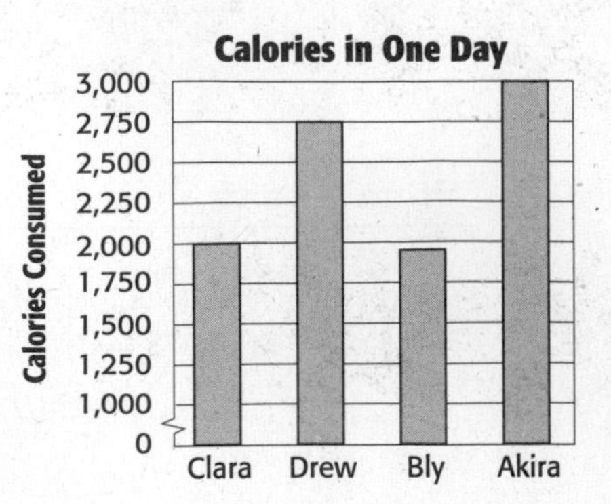

5. What does each bar represent?

6. Determine whether the graph is a bar graph or a histogram. Explain how you know.

8-5 Skills Practice

Problem-Solving Investigation: Use A Graph

Use a graph to solve the problem. For Exercises 1–3, refer to the graph.

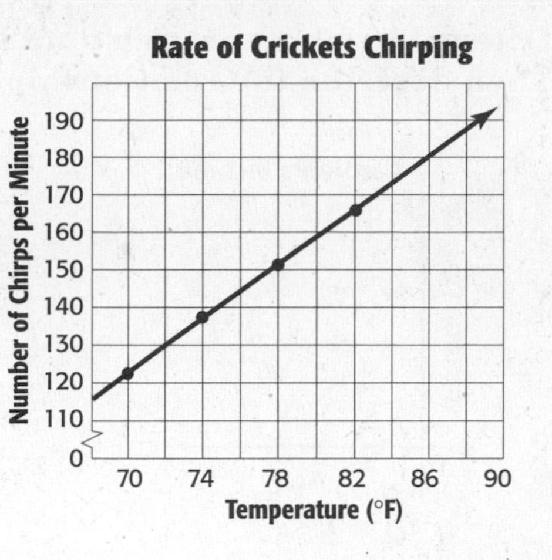

1. Estimate the temperature at which the rate of chirping is 130 per minute.

2. Predict the number of cricket chirps per minute at 86 degrees.

3. Predict the number of chirps per minute at 90 degrees.

For Exercises 4–6, refer to the graph.

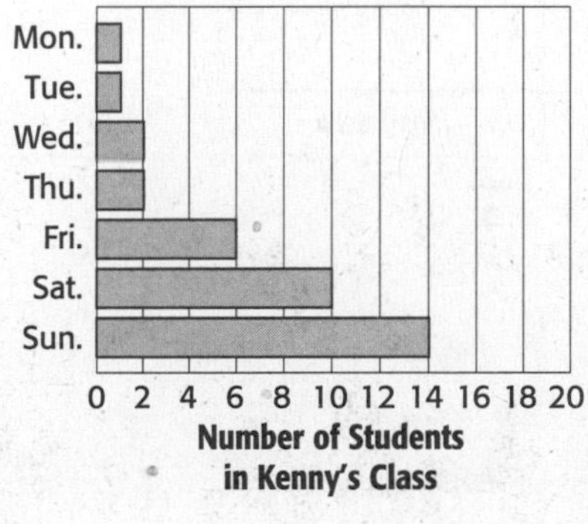

4. How many students consider Friday their favorite day of the week?

5. How many students prefer the weekend days?

6. How many students are in Kenny's class?

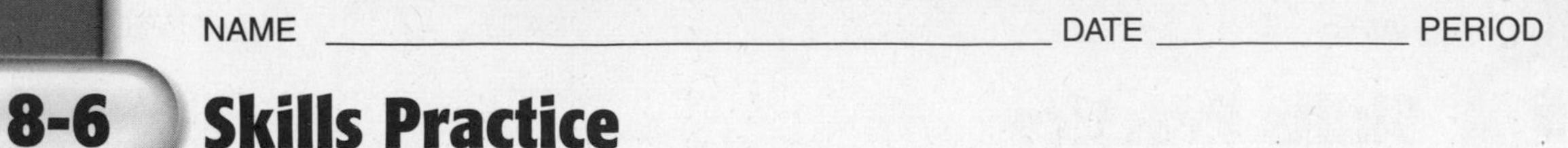

8-6 Skills Practice

Using Graphs to Predict

Determine whether each data set shows a *positive,* a *negative,* or *no* relationship. Then describe the relationship between the data sets.

1. **2.**

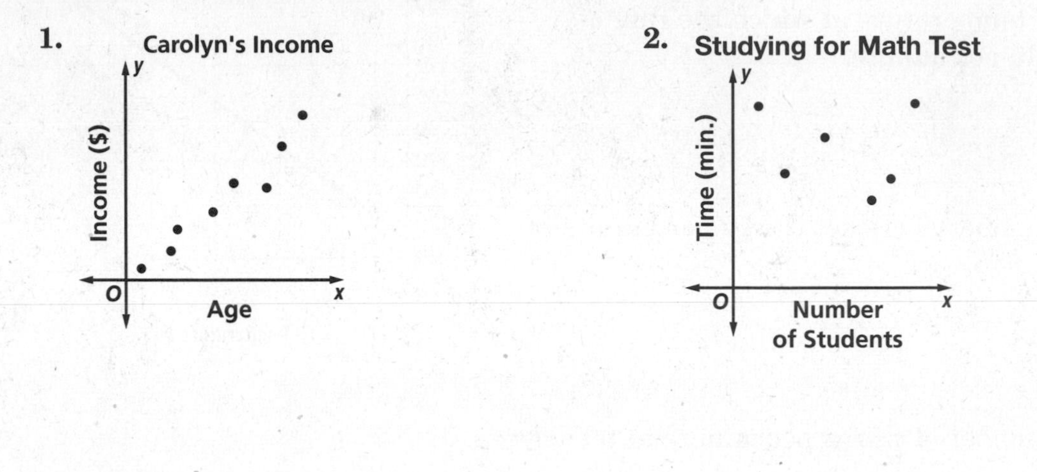

3.

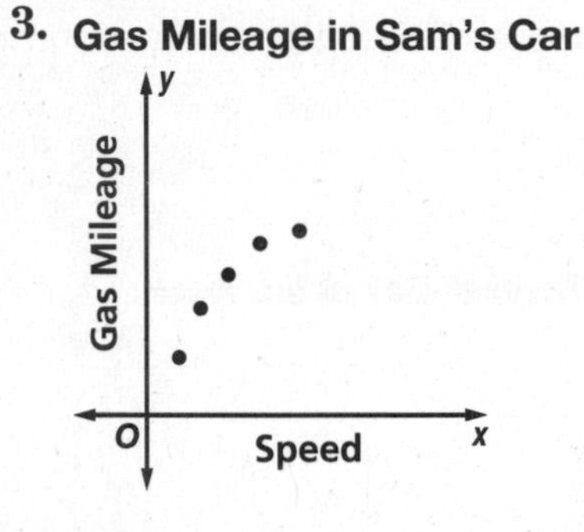

4.

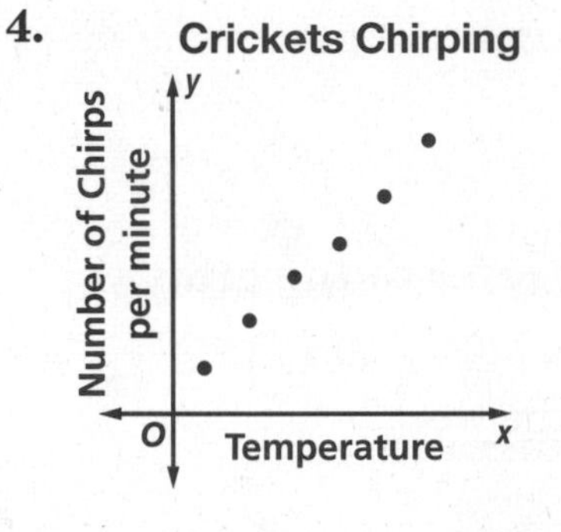

5.

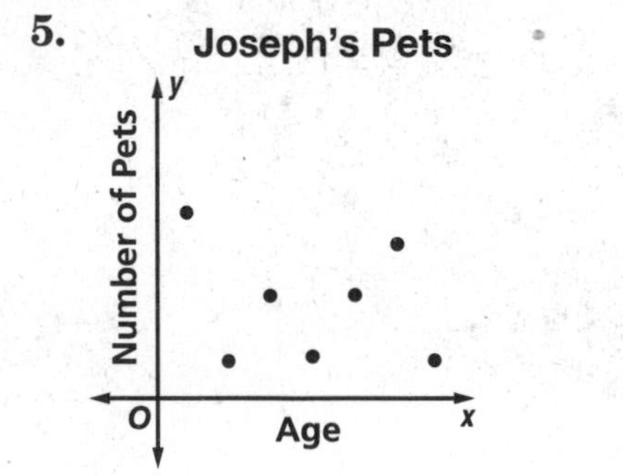

6.

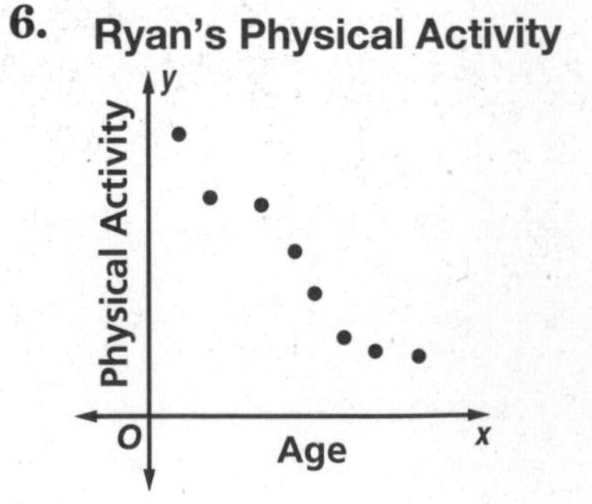

8-7 Skills Practice

Using Data to Predict

CELL PHONES For Exercises 1–3, use the table at the right. It shows the results of a survey in which students 12 to 17 years old were asked how often they use a cell phone.

Frequency of Use	Percent
more than twice a week	32%
once or twice a week	16%
once or twice a month	23%
less than once a month	12%
never used one	17%

1. Out of 215 students 12 to 17 years old, how many would you predict use a cell phone once or twice a week?

2. Predict how many students 12 to 17 years old in a group of 375 have never used a cell phone.

3. How many students 12 to 17 years old out of 1,200 would you expect use a cell phone at least once or twice a week?

PIZZA For Exercises 4–6, use the table at the right. It shows the results of a survey in which a random sample of seventh graders at Kiewit Middle School were asked to name their favorite pizza topping.

Pizza Topping	Percent
pepperoni	46%
peppers	28%
olives	8%
onions	2%
pineapple	4%
mushrooms	12%

4. There are 32 students in Mrs. Chen's seventh grade class. Predict how many would choose olives as their favorite topping.

5. There are 210 seventh grade students eating lunch in the cafeteria. How many of them would choose peppers as their favorite topping?

6. Predict how many of the 524 seventh graders at Kiewit Middle School would choose pepperoni as their favorite pizza topping.

7. BACKPACKS A survey showed that 78% of students who take a bus to school carry a backpack. Predict how many of the 654 students who take a bus also carry a backpack.

8-8 Skills Practice

Using Sampling to Predict

Each word in the box is a vocabulary word from lesson 8-8. Use the words to complete the sentences below. Not all of the words will be used.

unbiased biased samping	voluntary response sample simple random sample	convenience sample valid

1. A ______________________ is when members of the population are selected because they are easily accessed.

2. The survey is considered ______________________ when the entire population is represented.

3. It is called a ______________________ when each person in the population has an equal chance to be selected.

4. If only some members of the population choose to participate in a survey then it should be called a ______________________.

5. A sample would be considered ______________________ if one or more parts of he population are favored.

6. A conclusion can only be considered ______________________ when the information comes from an unbiased sample.

8-9 Skills Practice

Misleading Statistics

1. **INCOME** The bar graphs below show the total U.S. national income (nonfarm). Which graph could be misleading? Explain.

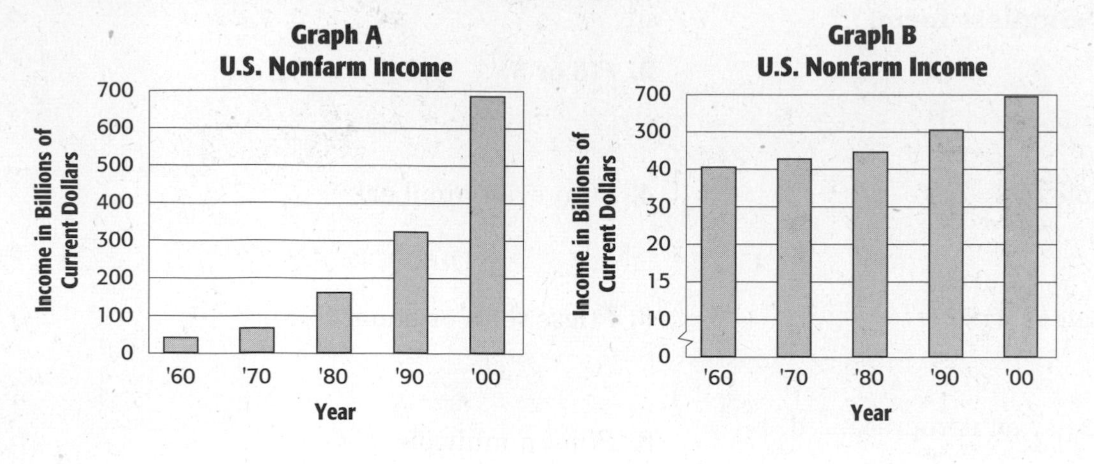

GEOGRAPHY For Exercises 2–4, use the table that shows the miles of shoreline for five states.

Miles of Shoreline	
State	**Length of Shoreline (mi)**
Virginia	3,315
Maryland	3,190
Washington	3,026
North Carolina	3,375
Pennsylvania	89

2. Find the mean, median, and mode of the data.

3. Which measure of central tendency is misleading in describing the miles of shoreline for the states? Explain.

4. Which measure of central tendency most accurately describes the data?

NAME ______________________ DATE ____________ PERIOD _____

9-1 Skills Practice

Simple Events

A set of 12 cards is numbered 1, 2, 3, ...12. Suppose you pick a card at random without looking. Find the probability of each event. Write as a fraction in simplest form.

1. *P*(5)

2. *P*(6 or 8)

3. *P*(a multiple of 3)

4. *P*(an even number)

5. *P*(a multiple of 4)

6. *P*(less than or equal to 8)

7. *P*(a factor of 12)

8. *P*(not a multiple of 4)

9. *P*(1, 3, or 11)

10. *P*(a multiple a 5)

The students at Job's high school were surveyed to determine their favorite foods. The results are shown in the table at the right. Suppose students were randomly selected and asked what their favorite food is. Find the probability of each event. Write as a fraction in simplest form.

Favorite Food	Responses
pizza	19
steak	8
chow mein	5
seafood	4
spaghetti	3
cereal	1

11. *P*(steak)

12. *P*(spaghetti)

13. *P*(cereal or seafood)

14. *P*(not chow mein)

15. *P*(pizza)

16. *P*(cereal or steak)

17. *P*(not steak)

18. *P*(not cereal or seafood)

19. *P*(chicken)

20. *P*(chow mein or spaghetti)

9-2 Skills Practice

Sample Spaces

The spinner at the right is spun twice.

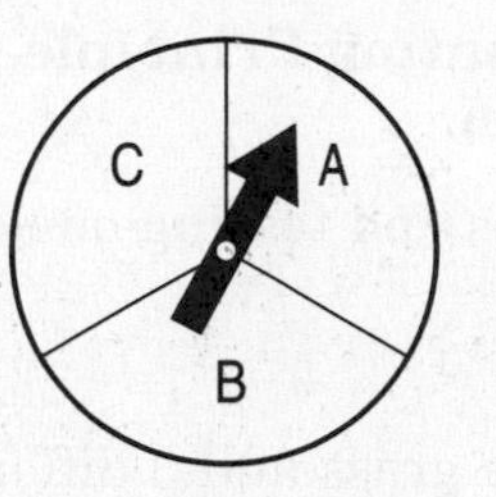

1. Draw a tree diagram to represent the situation.

2. What is the probability of getting at least one A?

For each situation, make a tree diagram or table to show the sample space. Then give the total number of outcomes.

3. choosing a hamburger or hot dog and potato salad or macaroni salad

4. choosing a vowel from the word COMPUTER and a consonant from the word BOOK

5. choosing between the numbers 1, 2 or 3, and the colors blue, red, or green

NAME ______________________________ DATE ____________ PERIOD _____

9-3 Skills Practice

The Fundamental Counting Principle

Use the Fundamental Counting Principle to find the total number of outcomes in each situation.

1. rolling two number cubes and tossing one coin

2. choosing rye or Bermuda grass and 3 different mixtures of fertilizer

3. making a sandwich with ham, turkey, or roast beef; Swiss or provolone cheese; and mustard or mayonaise

4. tossing 4 coins

5. choosing from 3 sizes of distilled, filtered, or spring water

6. choosing from 3 flavors of juice and 3 sizes

7. choosing from 35 flavors of ice cream; one, two, or three scoops; and sugar or waffle cone

8. picking a day of the week and a date in the month of April

9. rolling 3 number cubes and tossing 2 coins

10. choosing a 4-letter password using only vowels

11. choosing a bicycle with or without shock absorbers; with or without lights; and 5 color choices

12. a license plate that has 3 numbers from 0 to 9 and 2 letters

NAME ______________________ DATE ____________ PERIOD _____

9-4 Skills Practice

Permutations

Find the value of each expression.

1. $2 \cdot 1$

2. $4 \cdot 3 \cdot 2 \cdot 1$

3. $3 \cdot 2 \cdot 1 \cdot 5 \cdot 4 \cdot 3 \cdot 2 \cdot 1$

4. $9 \cdot 8 \cdot 7 \cdot 6 \cdot 5 \cdot 4 \cdot 3 \cdot 2 \cdot 1$

5. $2 \cdot 1 \cdot 8 \cdot 7 \cdot 6 \cdot 5 \cdot 4 \cdot 3 \cdot 2 \cdot 1$

6. $3 \cdot 2 \cdot 1 \cdot 2 \cdot 1$

7. $11 \cdot 10 \cdot 9$

8. $10 \cdot 9 \cdot 8 \cdot 7 \cdot 6 \cdot 5 \cdot 4 \cdot 3 \cdot 2 \cdot 1$

9. $5 \cdot 4 \cdot 3 \cdot 2 \cdot 1 \cdot 2 \cdot 1$

10. $5 \cdot 4 \cdot 3 \cdot 2$

11. $8 \cdot 7 \cdot 6 \cdot 5 \cdot 4 \cdot 3 \cdot 2 \cdot 1$

12. $6 \cdot 5 \cdot 4 \cdot 3 \cdot 2 \cdot 1$

13. How many ways can you arrange the letters in the word PRIME?

14. How many ways can you arrange 8 different crates on a shelf if they are placed from left to right?

Lesson 9-4

9-5 Skills Practice

Combinations

Tell whether each situation represents a *permutation* or *combination*. Then solve the problem.

1. You are allowed to omit two out of 12 questions on a quiz. How many ways can you select the questions to omit?

2. Six students are to be chosen from a class of 18 to represent the class at a math contest. How many ways can the six students be chosen?

3. How many different 5-digit zip codes are possible if no digits are repeated?

4. In a race with six runners, how many ways can the runners finish first, second, or third?

5. How many ways can two names be chosen from 76 in a raffle if only one entry per person is allowed?

6. How many ways can six students be arranged in a lunch line?

7. A family has a bike rack that fits seven bikes but they only have five bikes. How many ways can the bikes fit in the bike rack?

8. How many ways can you select three sheriff deputies from eight candidates?

9. How many ways can four finalists be selected from 50 contestants?

10. How many 4-digit pin numbers are available if no number is repeated?

11. How many handshakes can occur between five people if everyone shakes hands?

9-6 Skills Practice

Problem-Solving Investigation: Act it Out

Use the act it out strategy to solve.

1. **SCHOOL** Determine whether rolling a 6-sided number cube is a good way to answer a 20-question multiple-choice test if there are six choices for each question. Justify your answer.

2. **GYMNASTICS** Five gymnasts are entered in a competition. Assuming that there are no ties, how many ways can first, second, and third places be awarded?

3. **LUNCH** How many ways can 3 friends sit together in three seats at lunch?

4. **SCHEDULE** How many different schedules can Sheila create if she has to take English, math, science, social studies, and art next semester. Assume that there is only one lunch period available.

5. **BAND CONCERTS** The band is having a holiday concert. In the first row, the first trumpet is always furthest to the right and the first trombone is always the furthest to the left. How many ways are there to arrange the other 4 people who need to sit in the front?

6. **TEAMS** Mr. D is picking teams for volleyball is gym by having the students count off by 2's. The 1's will be on one team and the 2's on the other. Would flipping a coin would work just as well to pick the teams? Justify your answer.

NAME ______________________ DATE ____________ PERIOD _____

9-7 Skills Practice

Theoretical and Experimental Probability

For Exercises 1–5, a number cube is rolled 50 times and the results are shown in the graph below.

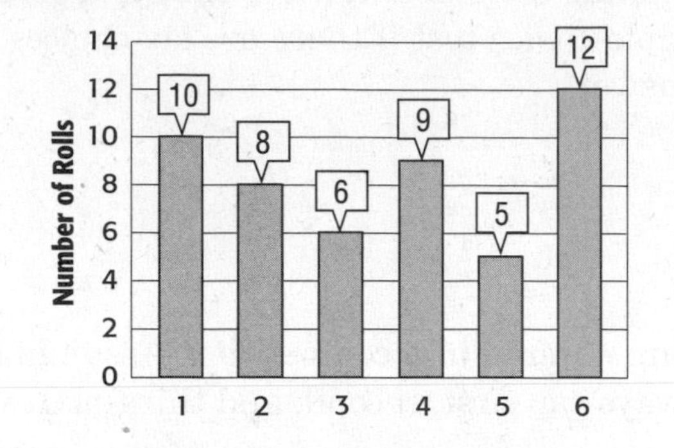

1. Find the experimental probability of rolling a 2.

2. What is the theoretical probability of rolling a 2?

3. Find the experimental probability of *not* rolling a 2.

4. What is the theoretical probability of *not* rolling a 2?

5. Find the experimental probability of rolling a 1.

For Exercises 6–9, use the results of the survey at the right.

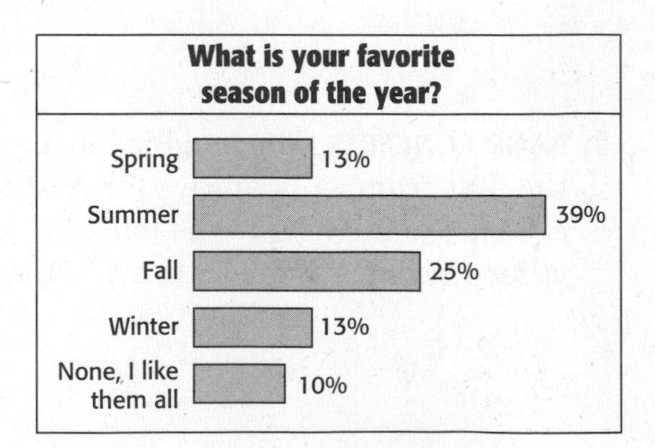

6. What is the probability that a person's favorite season is fall? Write the probability as a fraction.

7. Out of 300 people, how many would you expect to say that fall is their favorite season?

8. Out of 20 people, how many would you expect to say that they like all the seasons?

9. Out of 650 people, how many more would you expect to say that they like summer than say that they like winter?

NAME ______________________ DATE ____________ PERIOD _____

9-8 Skills Practice

Compound Events

1. Four coins are tossed. What is the probability of tossing all heads?

2. One letter is randomly selected from the word PRIME and one letter is randomly selected from the word MATH. What is the probability that both letters selected are vowels?

3. A card is chosen at random from a deck of 52 cards. It is then replaced and a second card is chosen. What is the probability of getting a jack and then an eight?

For Exercises 4–6, use the information below.

A standard deck of playing cards contains 52 cards in four suits of 13 cards each. Two suits are red and two suits are black. Find each probability. Assume the first card is replaced before the second card is drawn.

4. *P*(black, queen)

5. *P*(black, diamond)

6. *P*(jack, queen)

7. What is the probability of spinning a number greater than 5 on a spinner numbered 1 to 8 and tossing a tail on a coin?

8. Two cards are chosen at random from a standard deck of cards with replacement. What is the probability of getting 2 aces?

9. A CD rack has 8 classical CDs, 5 pop CDs, and 3 rock CDs. One CD is chosen and replaced, then a second CD is chosen. What is the probability of choosing a rock CD then a classical CD?

10. A jar holds 15 red pencils and 10 blue pencils. What is the probability of drawing one red pencil from the jar?

NAME ______________________ DATE ____________ PERIOD _____

10-1 Skills Practice

Angle Relationships

Classify each angle as *acute*, *obtuse*, *right*, or *straight*.

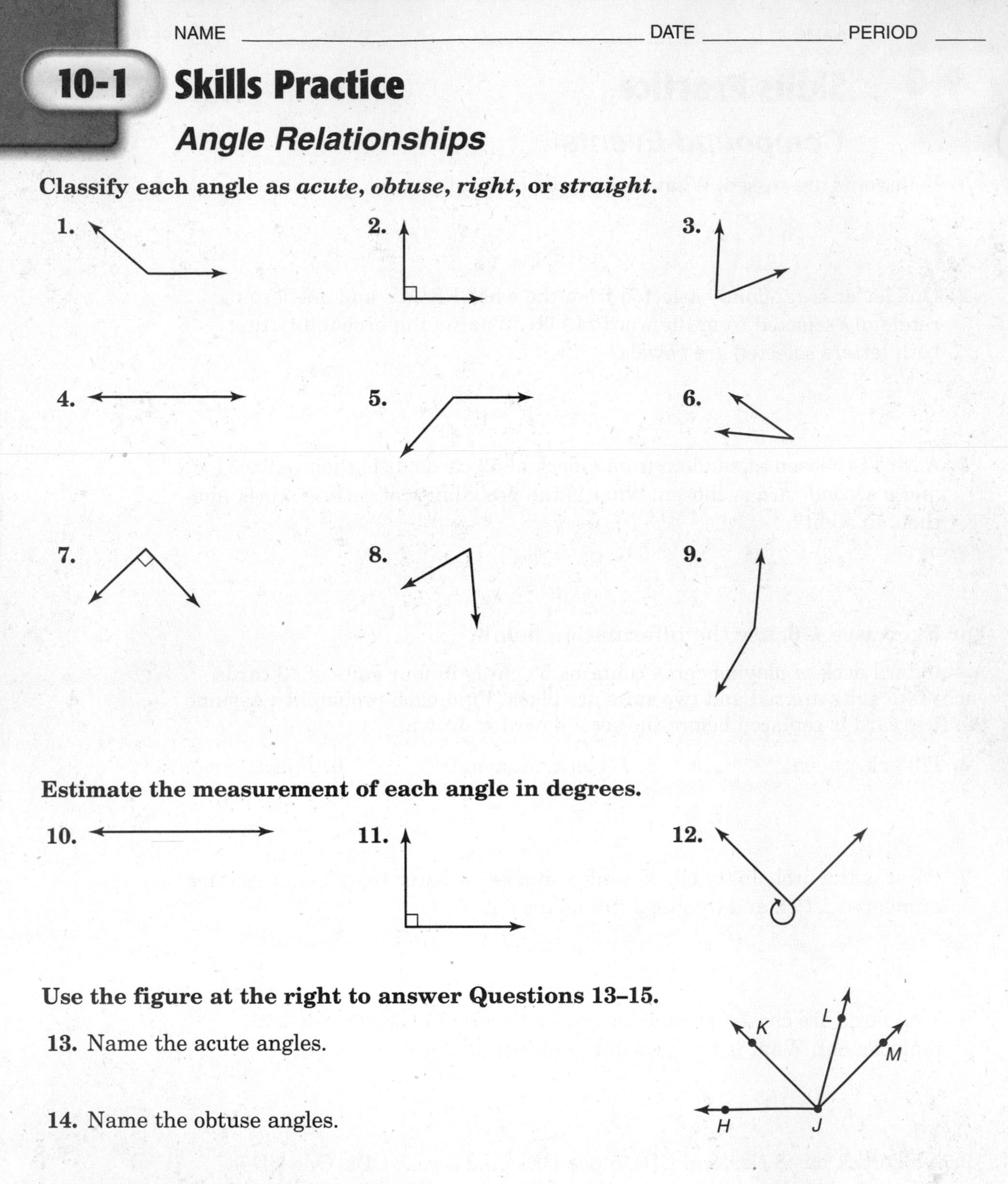

1.

2.

3.

4.

5.

6.

7.

8.

9.

Estimate the measurement of each angle in degrees.

10.

11.

12.

Use the figure at the right to answer Questions 13–15.

13. Name the acute angles.

14. Name the obtuse angles.

15. Name two angles that are adjacent.

NAME _______________ DATE _______ PERIOD _____

10-2 Skills Practice

Complementary and Supplementary Angles

Find the missing angle measure.

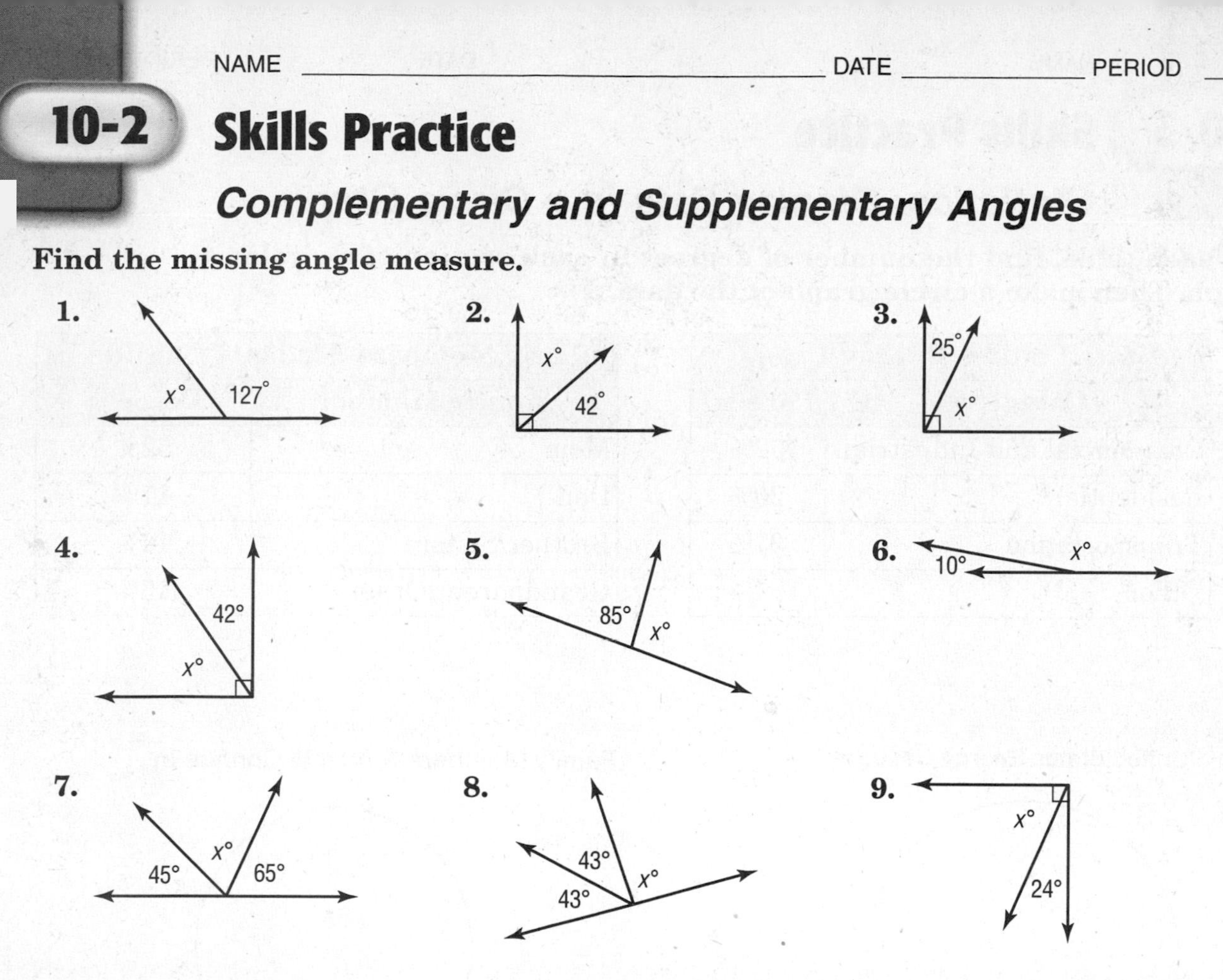

Use the figure at the right to answer Questions 10–13.

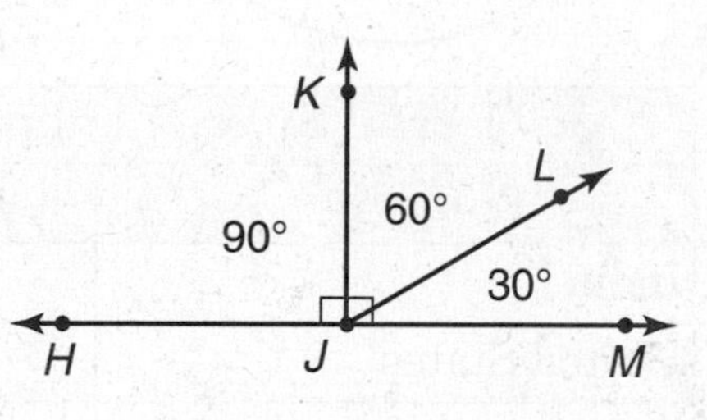

10. Find $m\angle HJL$.

11. $\angle HJL$ and $\angle LJM$ are what type of angles?

12. Find $m\angle KJM$.

13. $\angle KJL$ and $\angle LJM$ are what type of angles?

Lesson 10-2

10-3 Skills Practice

Statistics: Display Data in a Circle Graph

For each table, find the number of degrees in each section of a circle graph. Then make a circle graph of the data.

1.

United States Energy Usage	
Category	**Percent**
Commercial and Industrial	52%
Residential	20%
Transportation	27%
Other	1%

2.

Family Members Students Confide In	
Family Member	**Percent**
Mom	52%
Dad	17%
Brother/Sister	16%
Grandparent/Other	15%

United States Energy Usage

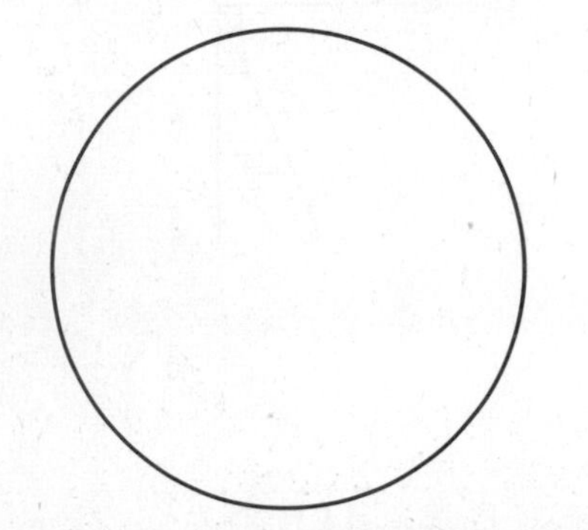

Family Members Students Confide In

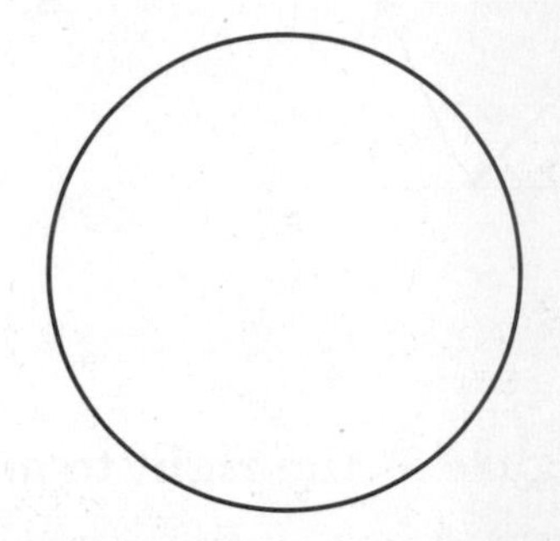

3.

Successful Space Launches	
Country	**Number**
India	2
United States	23
European Space Agency	7
China	1

4.

United States Coastline	
Coast	**Length (mi)**
Atlantic	2,100
Pacific	7,600
Gulf	1,600
Arctic	1,100

Successful Space Launches

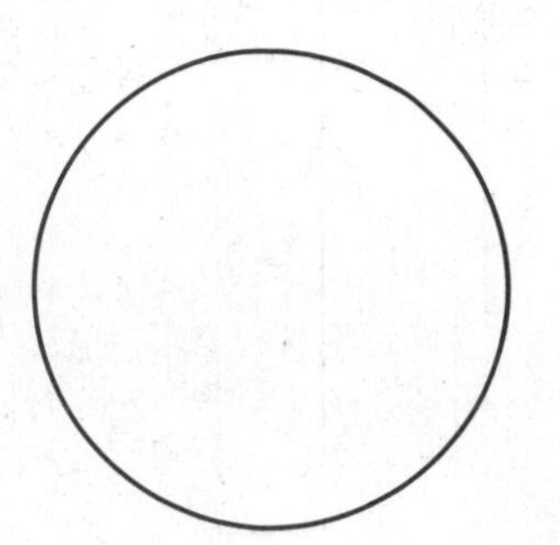

United States Coastline

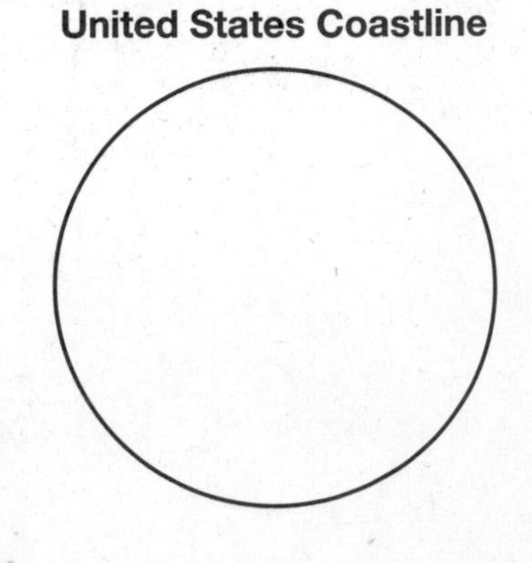

NAME ______________________ DATE ____________ PERIOD _____

10-4 Skills Practice

Triangles

Find the missing measure in each triangle. Then classify the triangle as *acute*, *right*, or *obtuse*.

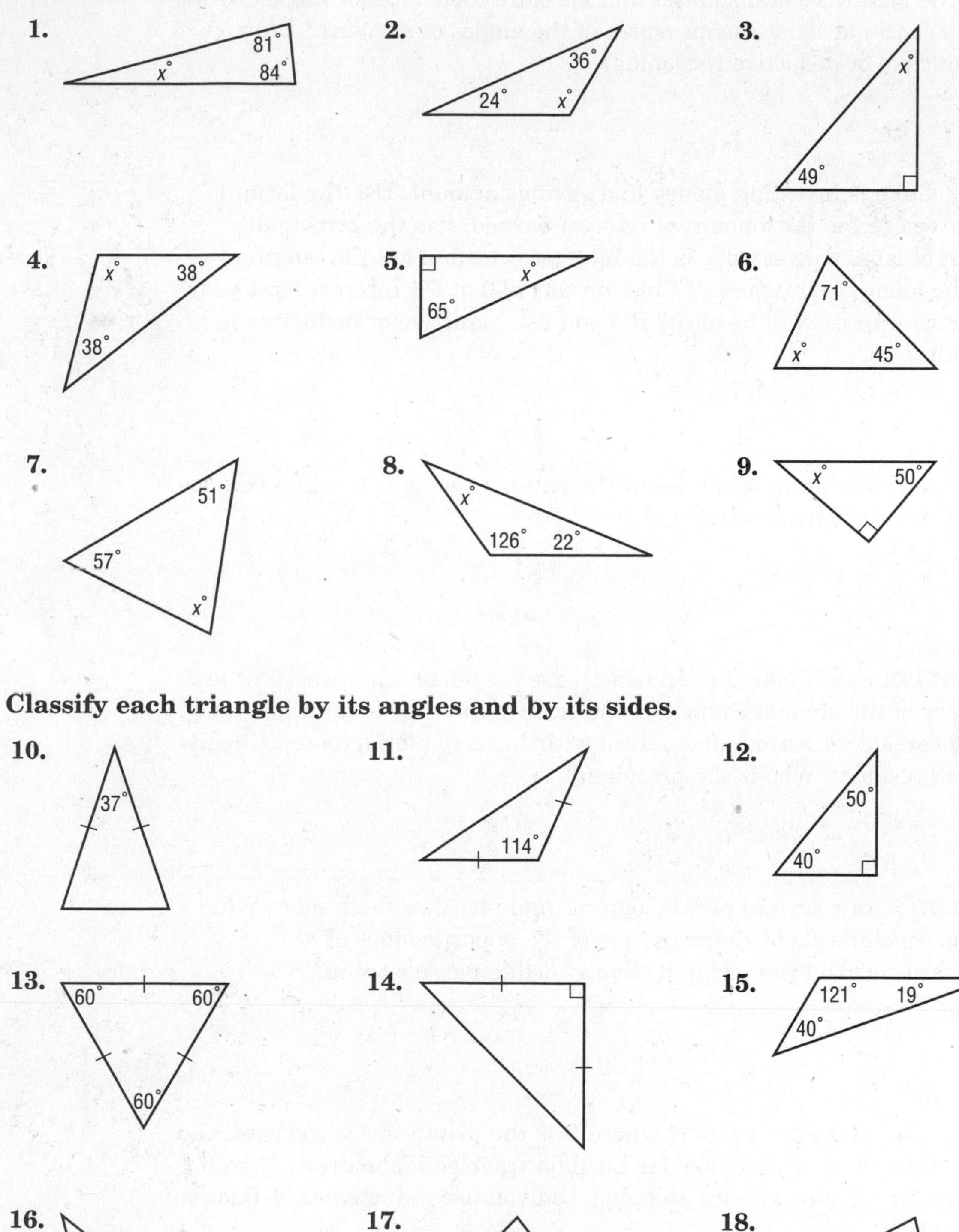

Classify each triangle by its angles and by its sides.

NAME ______________________________ DATE ____________ PERIOD _____

10-5 Skills Practice

Problem-Solving Investigation: Use Logical Reasoning

Use logical reasoning to solve.

1. **GEOMETRY** Draw several squares and measure their interior angles. What can you conclude about the measures of the angles of a square? Did you use inductive or deductive reasoning?

2. **MONEY** Luke is investing money in a savings account. Use the formula $I = Prt$ where I is the amount of interest earned, P is the principal amount of money invested, r is the interest rate, and t is the length of time the money is invested. If Luke invests 500 at 5% interest for 1 year, how much interest will he earn? Did you use inductive or deductive reasoning?

3. **PATTERNS** Write a rule to represent the pattern shown below. Did you use inductive or deductive reasoning?
   2, 4, 6, 8, 10, …

4. **STUDENT COUNCIL** Chen, Sue and Jacob are president, vice president and secretary of the student council, not necessarily in that order. Chen and the vice president stayed after school with Jacob to plan a dance. Chen is not the president. Who is the president?

5. **GEOMETRY** Draw several parallelograms and measure their sides. What can you conclude about the measures of the opposite sides of a parallelogram? Did you use inductive or deductive reasoning?

6. **TRAVEL** Use the formula $D = rt$ where D is the distance, r is the rate, and t is the time to determine how far Lucinda traveled if she drove 65 miles per hour for 6 hours without stopping. Did you use inductive or deductive reasoning?

NAME ______________________ DATE ____________ PERIOD _____

10-6 Skills Practice

Quadrilaterals

Classify the quadrilateral using the name that *best* describes it.

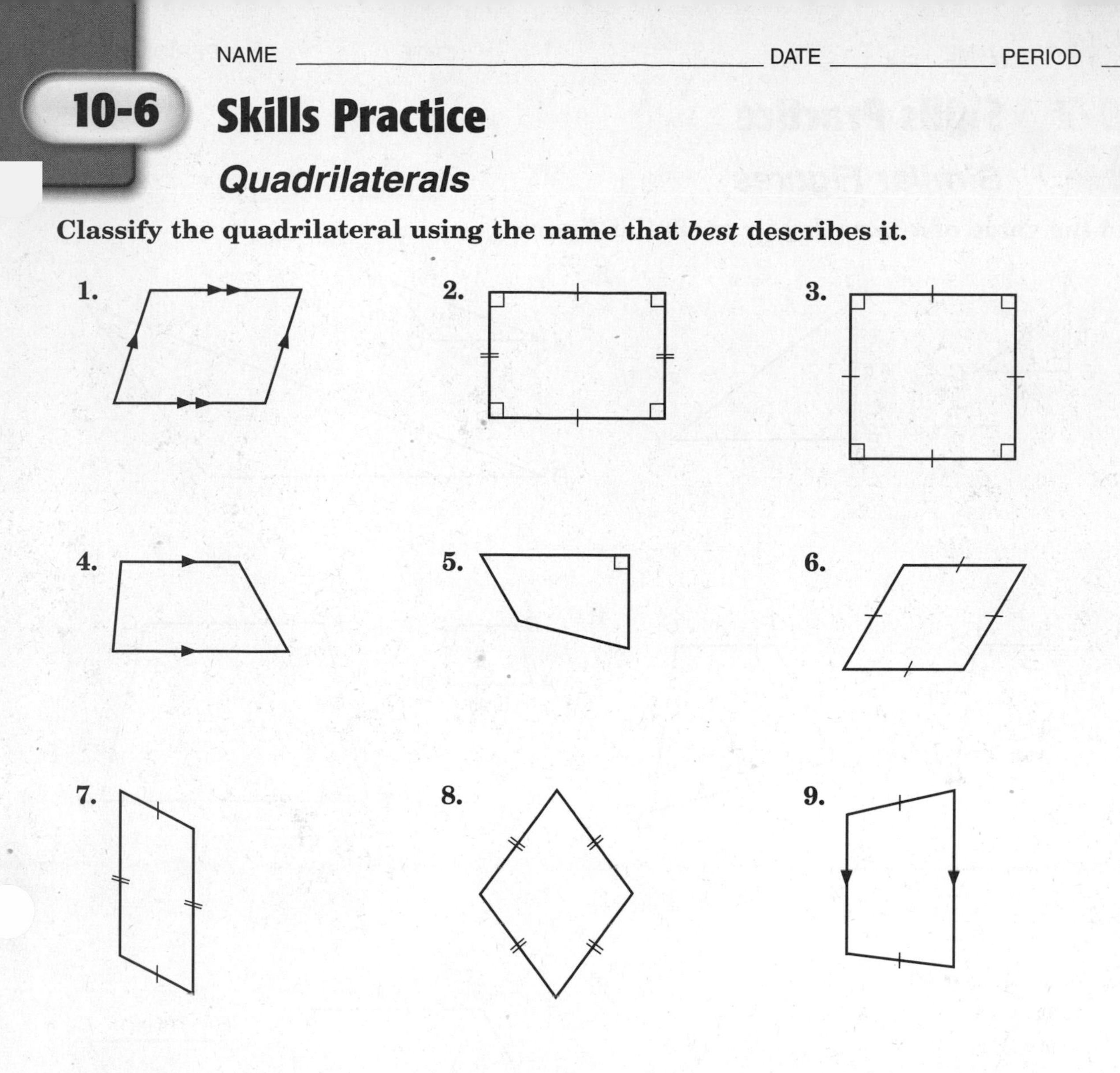

Find the missing angle measure of each quadrilateral.

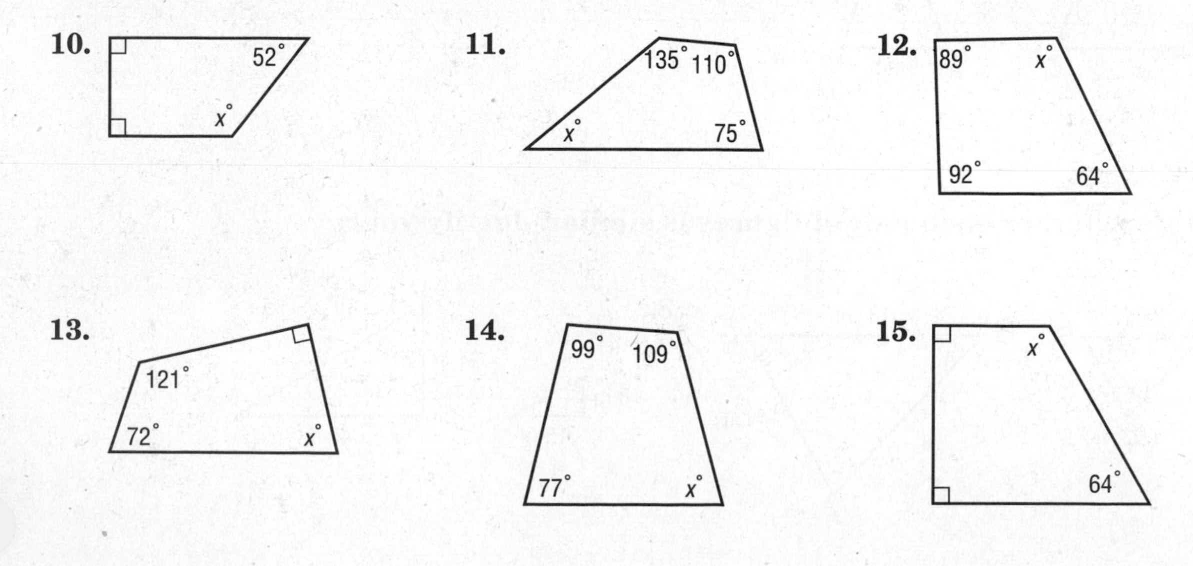

Lesson 10-6

NAME ______________________ DATE ____________ PERIOD _____

10-7 Skills Practice

Similar Figures

Find the value of *x* in each pair of similar figures.

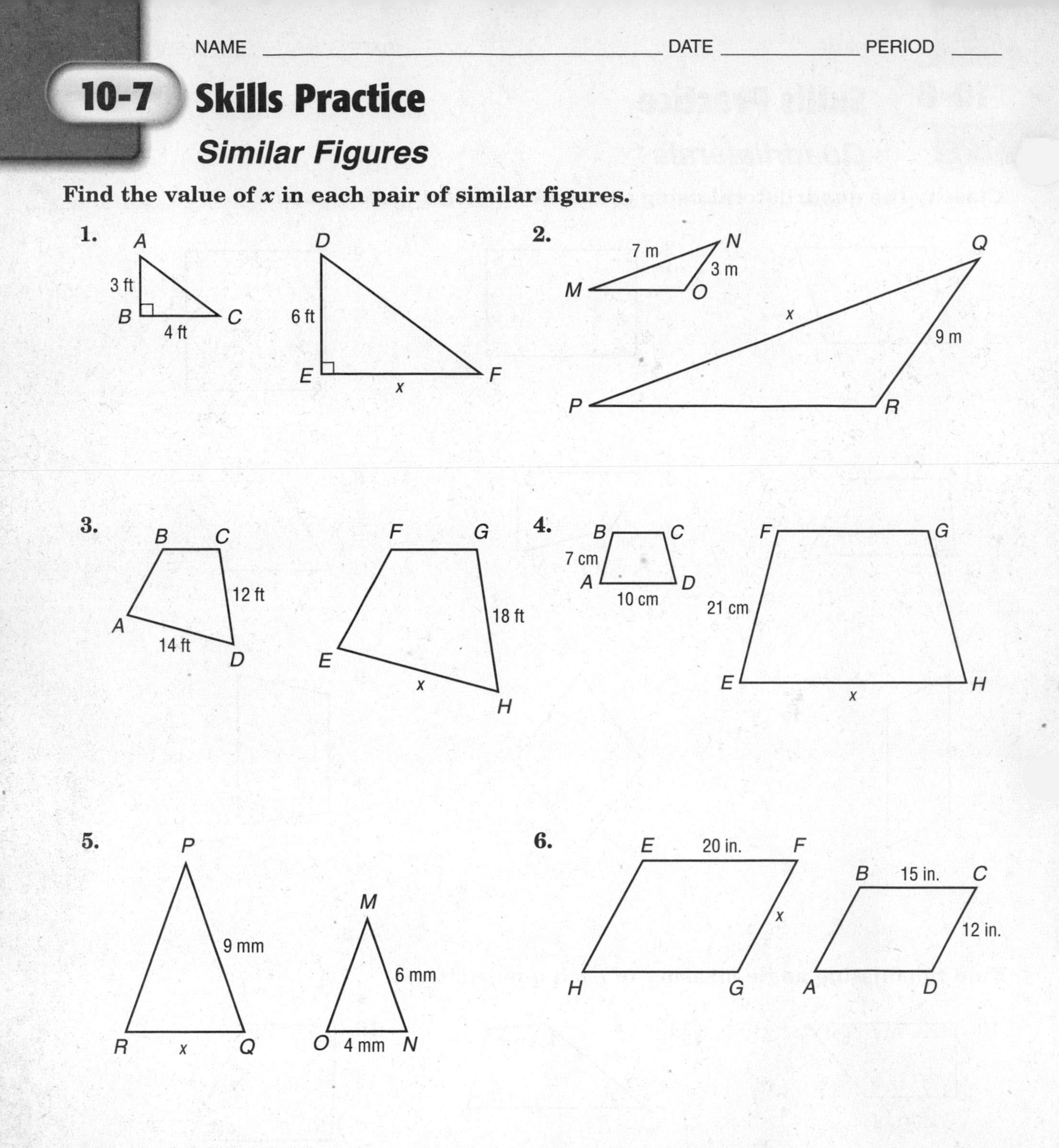

Determine whether each pair of figures is similar. Justify your answer.

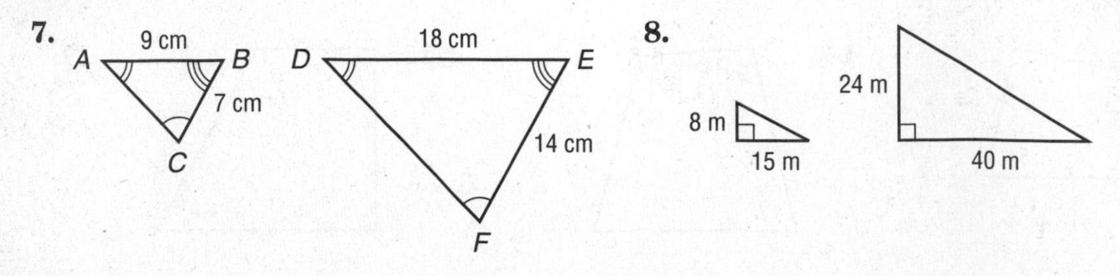

10-8 Skills Practice

Polygons and Tessellations

Determine whether each figure is a polygon. If it is, classify the polygon and state whether it is regular. If it is *not* a polygon, explain why.

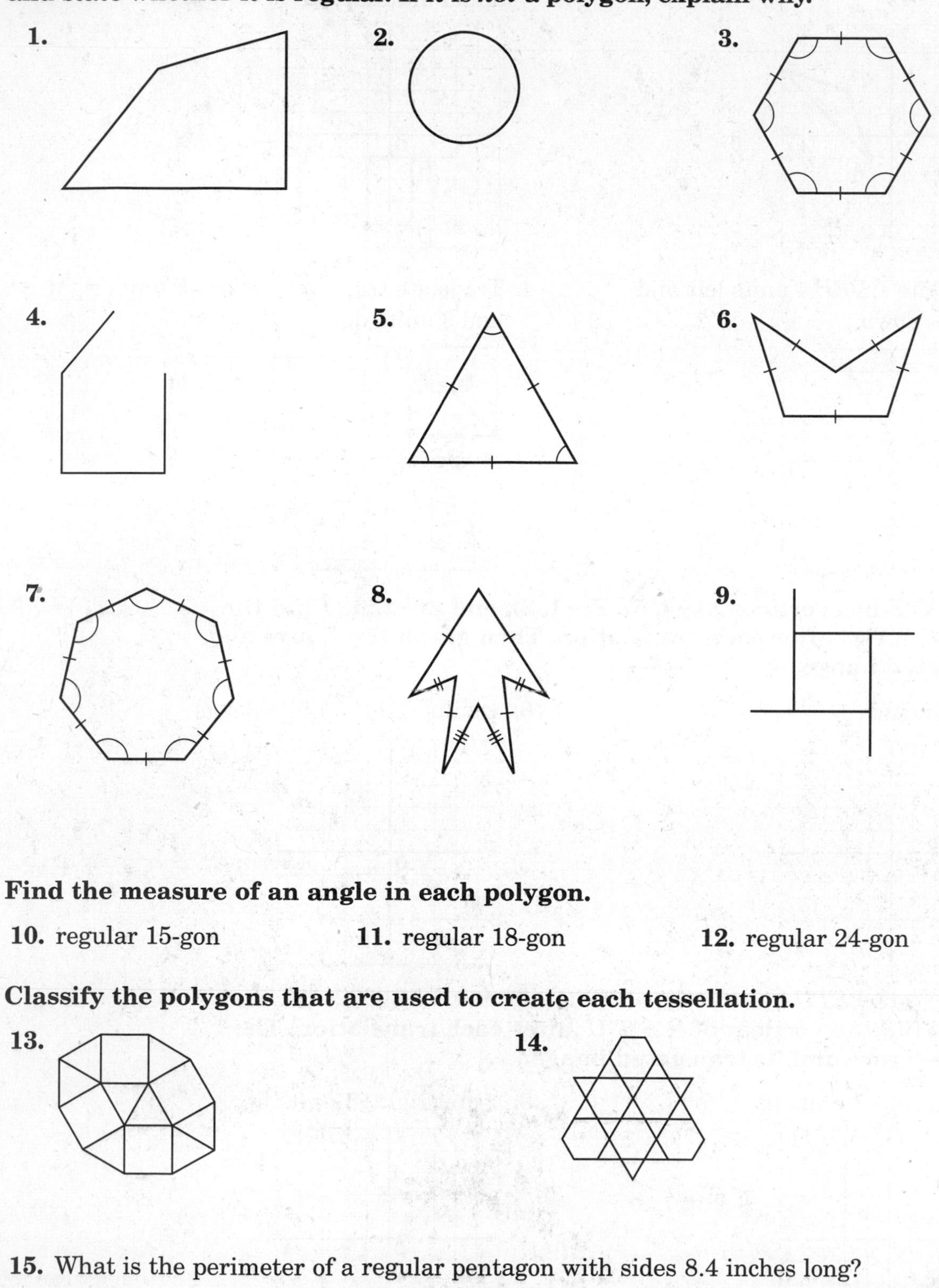

Find the measure of an angle in each polygon.

10. regular 15-gon

11. regular 18-gon

12. regular 24-gon

Classify the polygons that are used to create each tessellation.

13.

14.

15. What is the perimeter of a regular pentagon with sides 8.4 inches long?

NAME ______________________ DATE ____________ PERIOD _____

10-9 Skills Practice

Translations

1. Translate $\triangle ABC$ 5 units left.

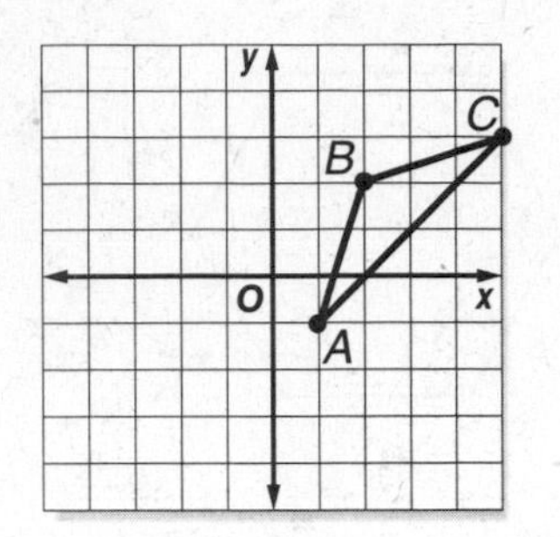

2. Translate rectangle *RSTU* 2 units right and 5 units up.

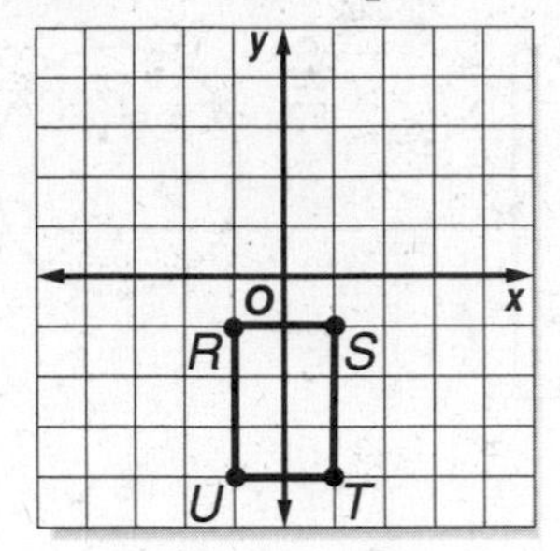

3. Translate $\triangle DEF$ 4 units left and 4 units down.

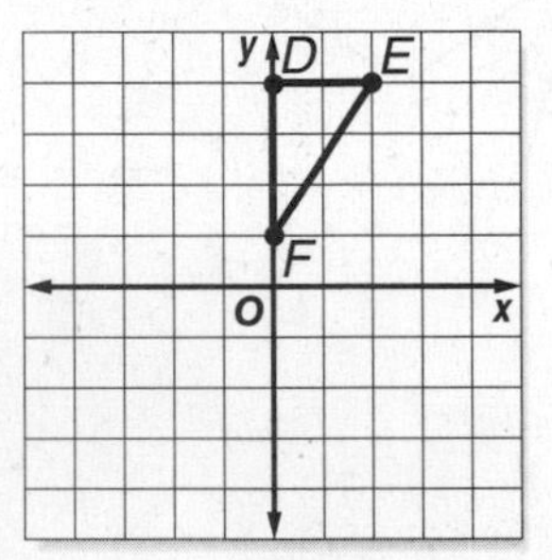

4. Translate trapezoid *LMNO* 5 units right and 3 units down.

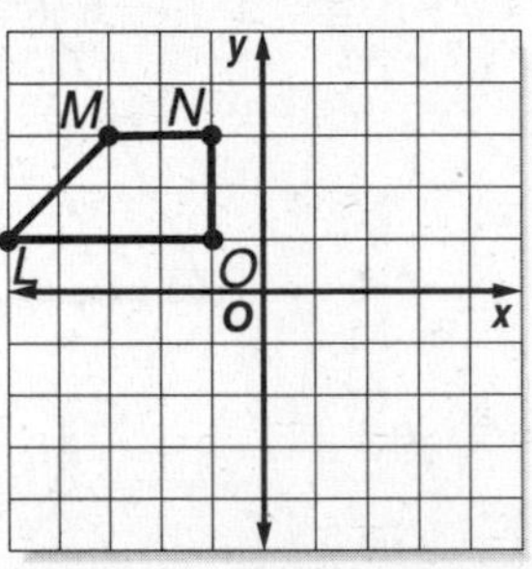

Triangle *XYZ* has vertices *X*(−4, 5), *Y*(−1, 3), and *Z*(−2, 0). Find the vertices of *X′Y′Z′* after each translation. Then graph the figure and its translated image.

5. 5 units down

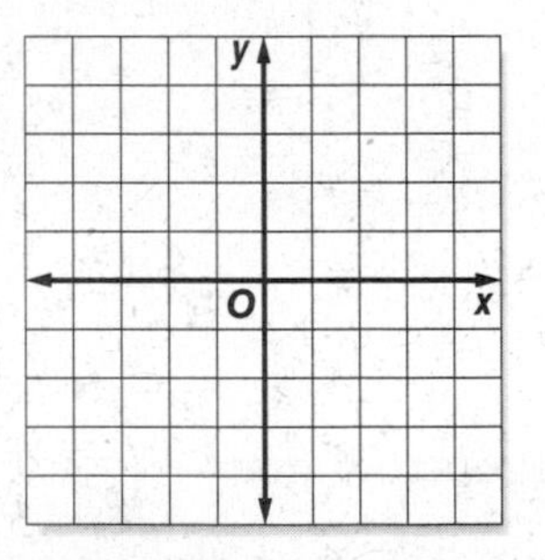

6. 4 units right, 3 units down

Parallelogram *RSTU* has vertices *R*(−1, −3), *S*(0, −1), *T*(4, −1), and *U*(3, −3). Find the vertices of *R′S′T′U′* after each translation. Then graph the figure and its translated image.

7. 3 units left, 3 units up

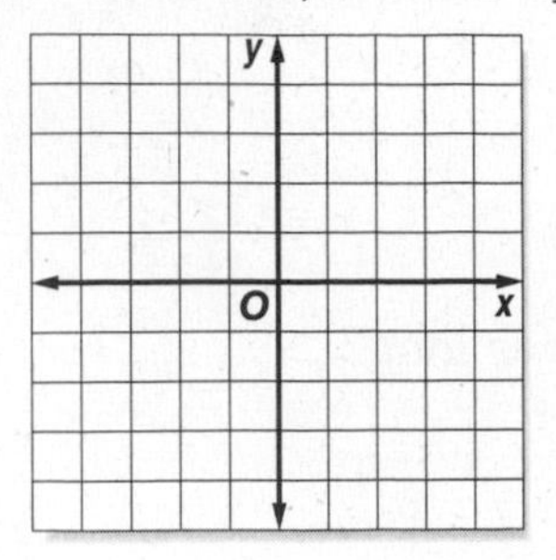

8. 1 unit right, 5 units up

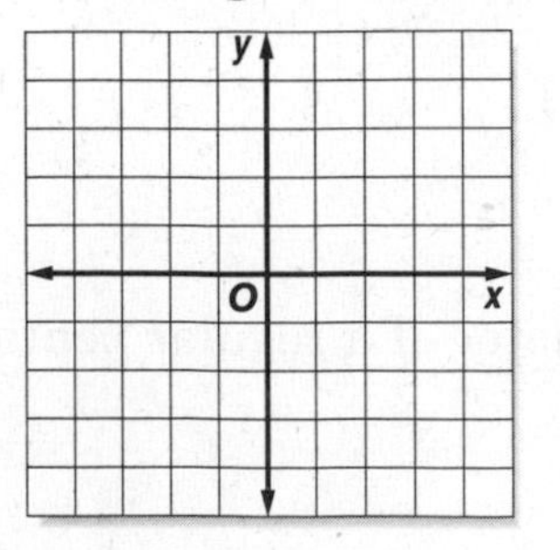

NAME ______________________ DATE ____________ PERIOD _____

10-10 Skills Practice

Reflections

Determine which figures have line symmetry. Then draw all lines of symmetry.

1. **2.** **3.**

4. **5.** **6.**

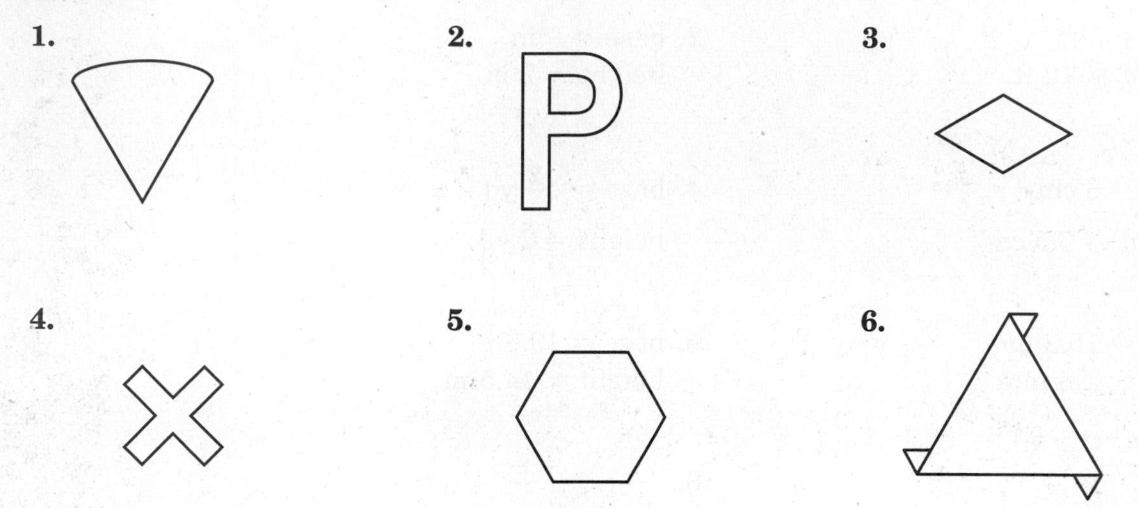

Graph the figure and its reflection over the *x*-axis. Then find the coordinates of the reflected image.

7. triangle ABC with vertices $A(-3, 4)$, $B(1, 4)$, and $C(3, 1)$

8. rectangle $MNOP$ with vertices $M(-2, -4)$, $N(-2, -1)$, $O(3, -1)$, and $P(3, -4)$

Graph the figure and its reflection over the *y*-axis. Then find the coordinates of the reflected image.

9. triangle DEF with vertices $D(1, 4)$, $E(4, 3)$, and $F(2, 0)$

10. trapezoid $WXYZ$ with vertices $W(-1, 3)$, $X(-1, -4)$, $Y(-5, -4)$, and $Z(-3, 3)$

11-1 Skills Practice

Area of Parallelograms

Find the area of each parallelogram. Round to the nearest tenth if necessary.

1. base = 5 ft
height = 12 ft

2. base = 9 in.
height = 2 in.

3. base = 6 cm
height = 5.5 cm

4. base = $4\frac{2}{5}$ yd
height = 2 yd

5. base = 15.3 mm
height = 8 mm

6. base = 19.6 m
height = 14.5 m

7.

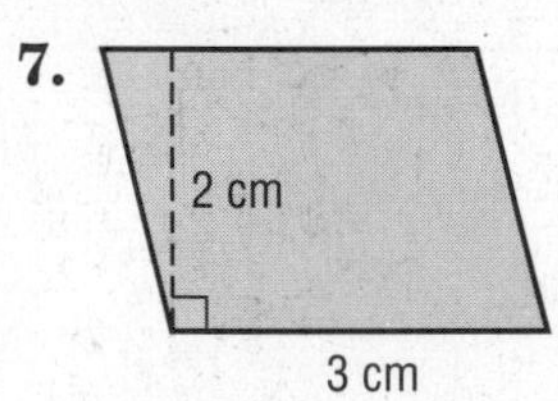

8.

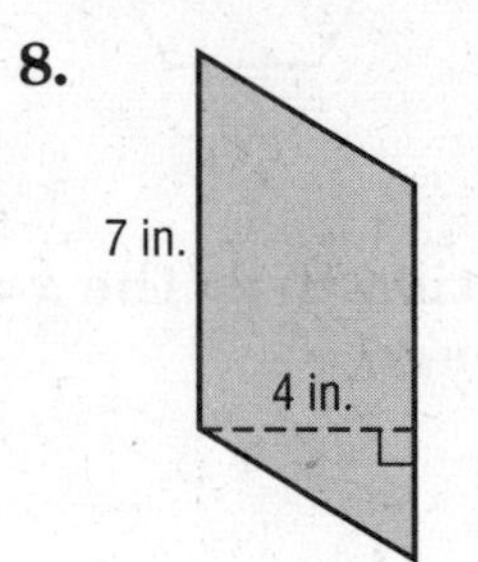

9.

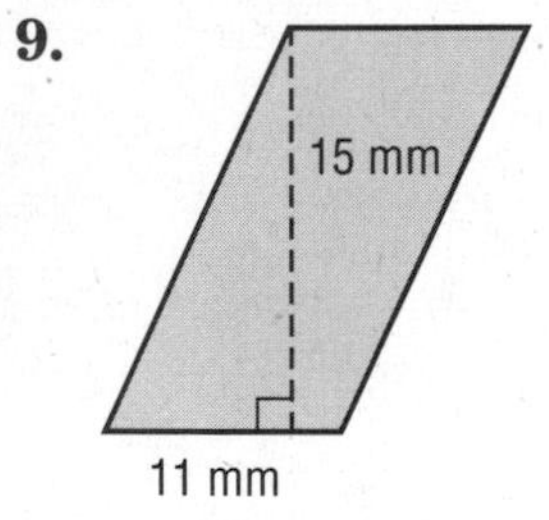

10.

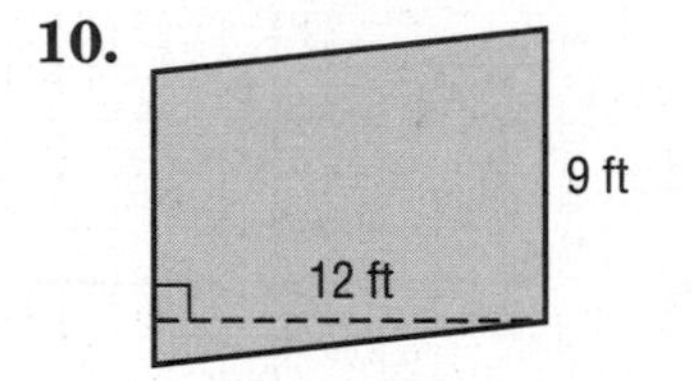

11.

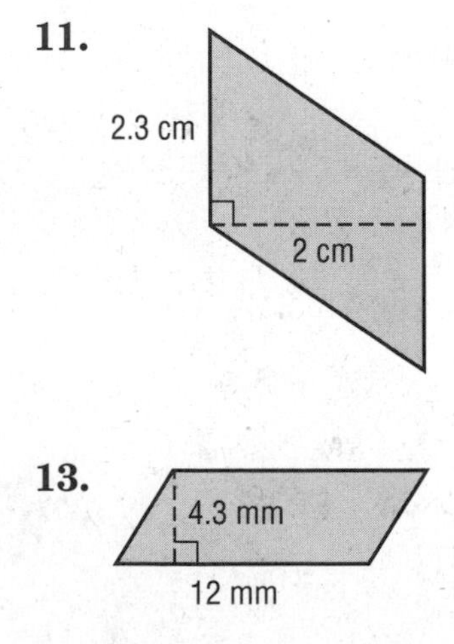

12.

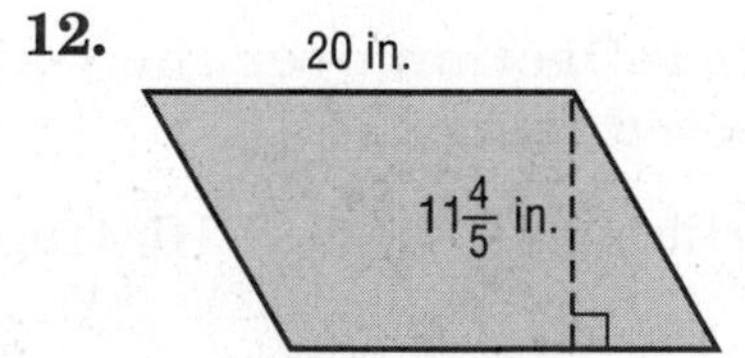

13.

14.

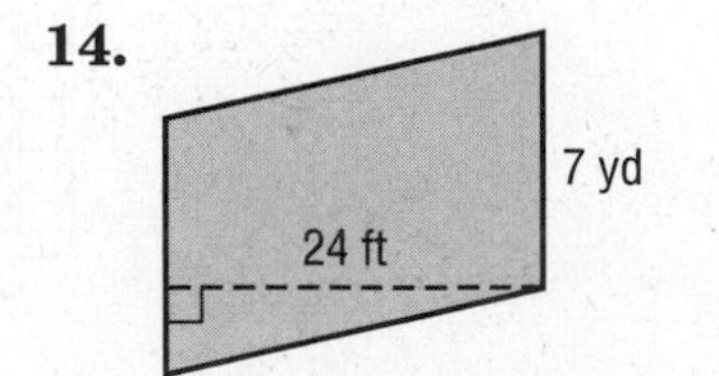

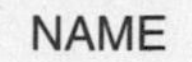

NAME ______________________ DATE __________ PERIOD _____

11-2 Skills Practice

Area of Triangles and Trapezoids

Find the area of each figure. Round to the nearest tenth if necessary.

1.

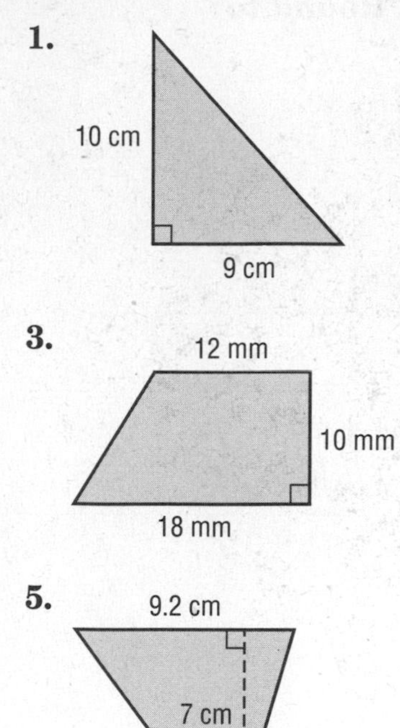

2.

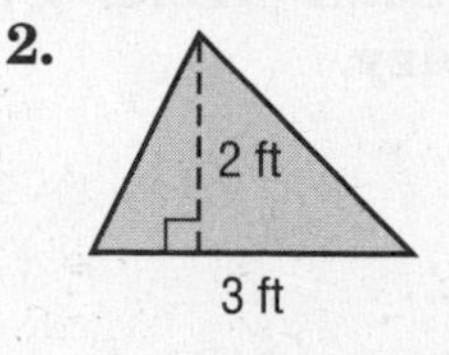

3.

4.

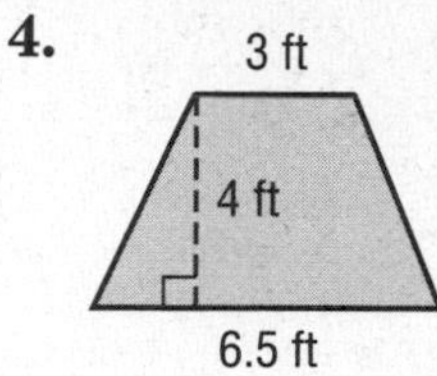

5.

9.2 cm

7 cm

2 cm

6.

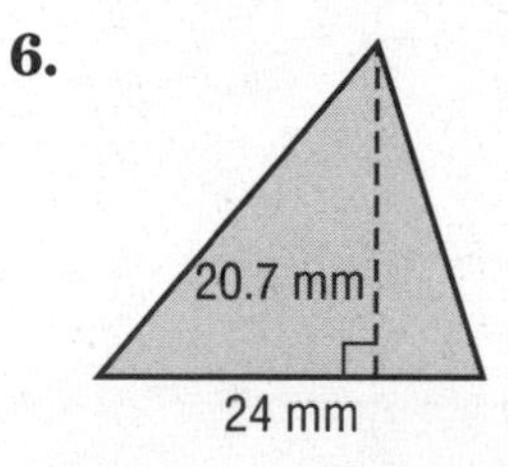

7.

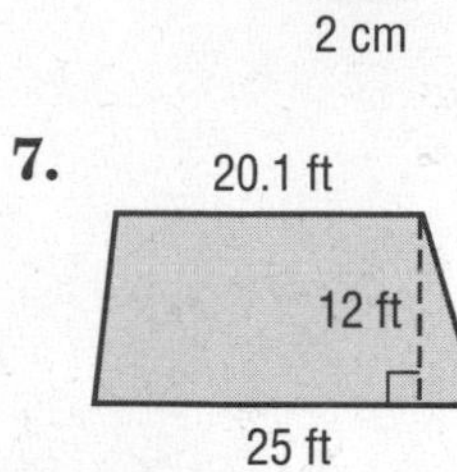

8.

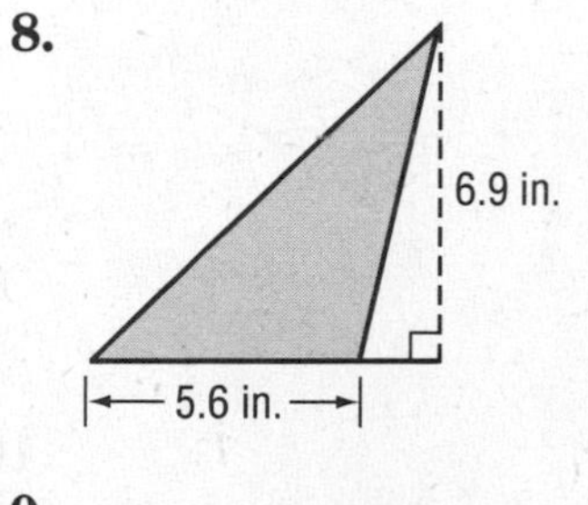

9.

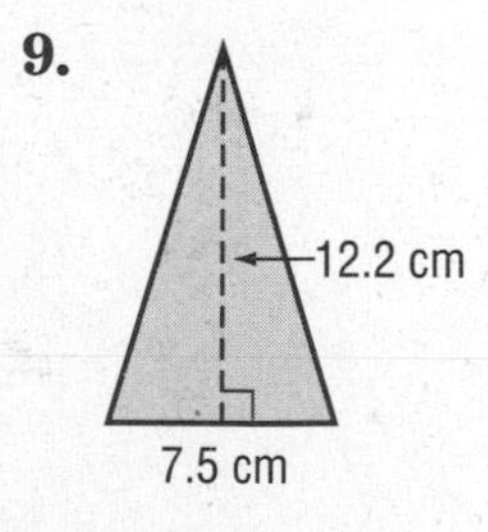

10.

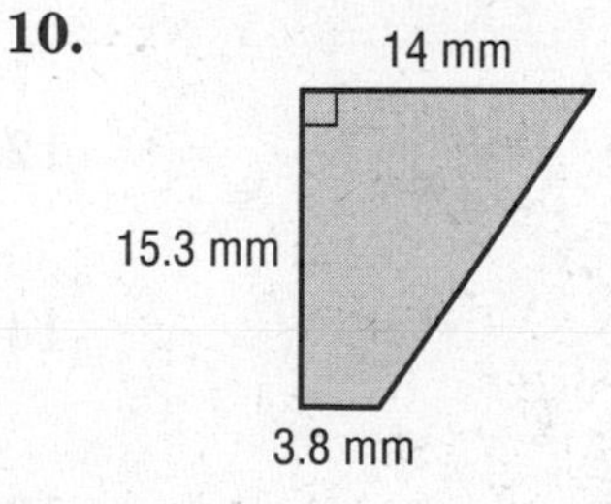

11. triangle: base = 16 cm, height = 9.4 cm

12. triangle: base = 13.5 in., height = 6.4 in.

13. trapezoid: bases 22.8 mm and 19.7 mm, height 36 mm

14. trapezoid: bases 5 ft and $3\frac{1}{2}$ ft, height 7 ft

Lesson 11-2

11-3 Skills Practice

Circles and Circumference

Find the circumference of each circle. Use 3.14 or $\frac{22}{7}$ for π. Round to the nearest tenth if necessary.

1.

3.

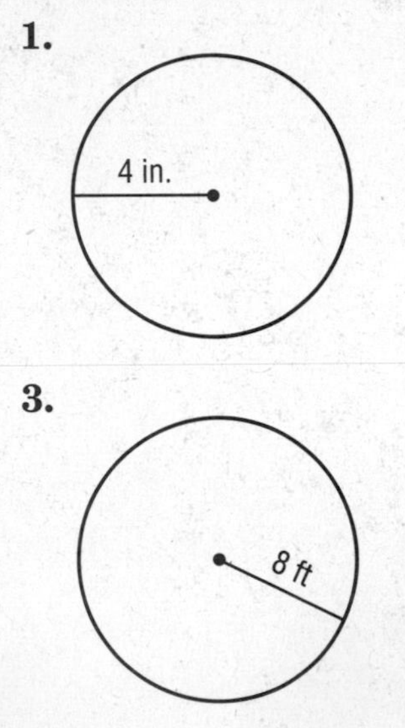

5.

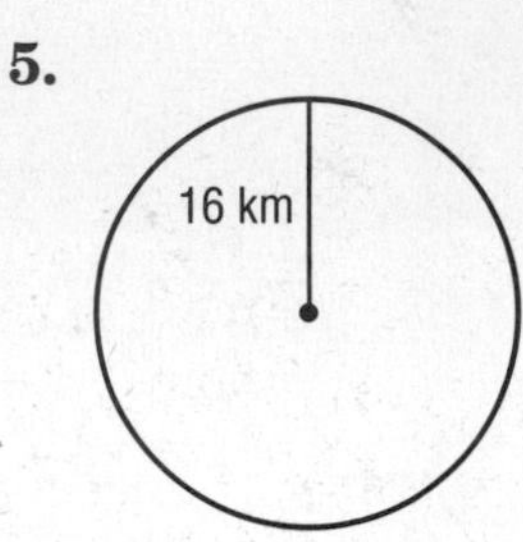

2.

4.

6.

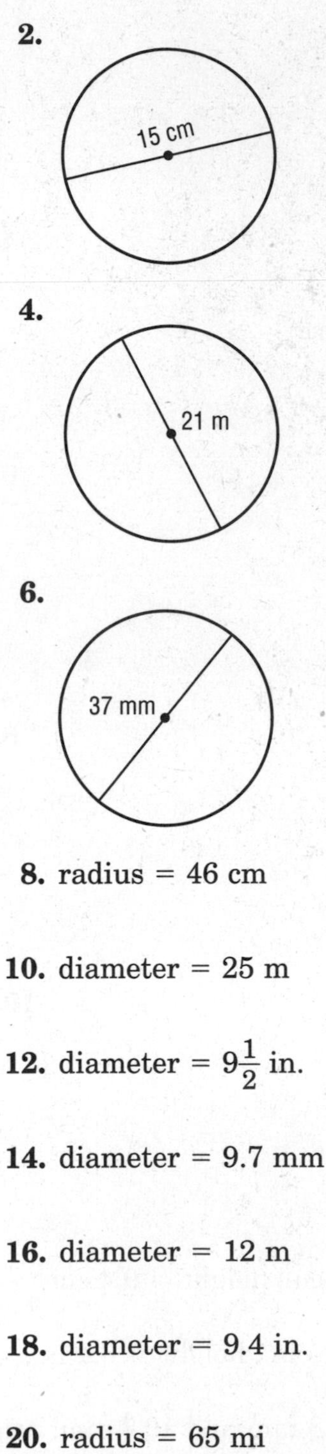

7. radius = 3 km

8. radius = 46 cm

9. diameter = 30 in.

10. diameter = 25 m

11. radius = 5 ft

12. diameter = $9\frac{1}{2}$ in.

13. radius = $3\frac{1}{2}$ ft

14. diameter = 9.7 mm

15. radius = 5.2 km

16. diameter = 12 m

17. radius = 22 ft

18. diameter = 9.4 in.

19. radius = 100 m

20. radius = 65 mi

21. diameter = $10\frac{1}{2}$ in.

22. diameter = 8.5 cm

NAME ______________________ DATE ____________ PERIOD _____

11-4 Skills Practice

Area of Circles

Find the area of each circle. Use 3.14 for π. Round to the nearest tenth.

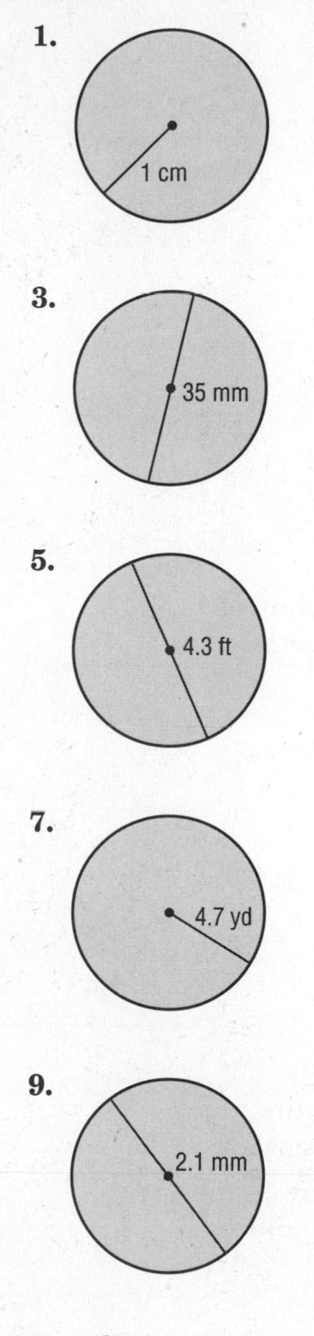

2.

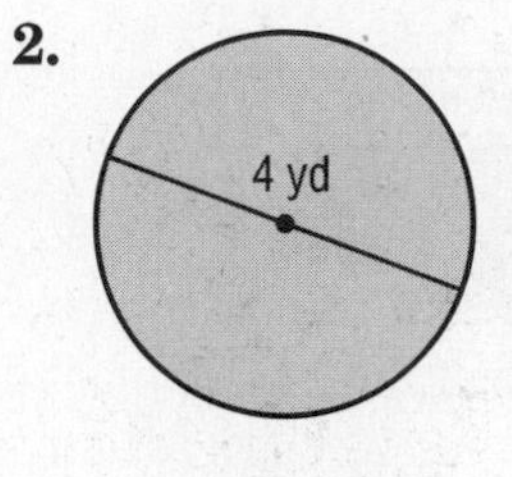

4.

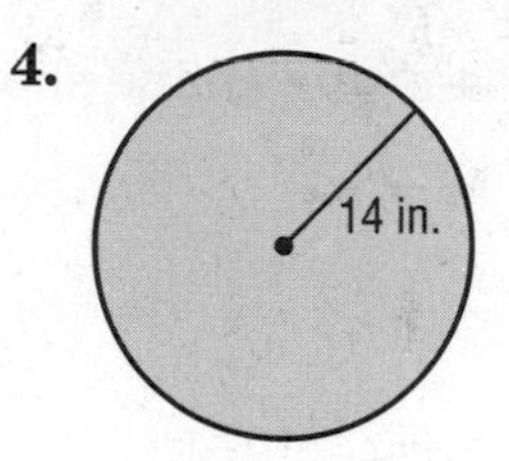

6.

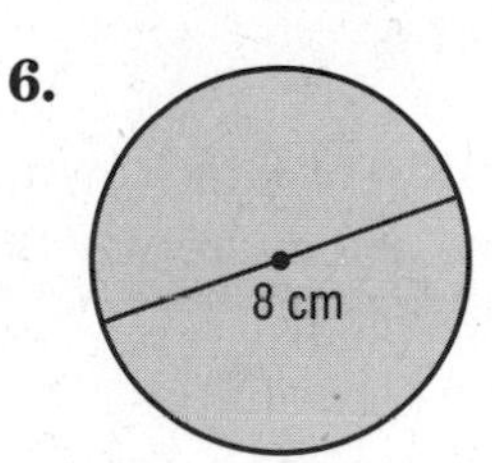

8.

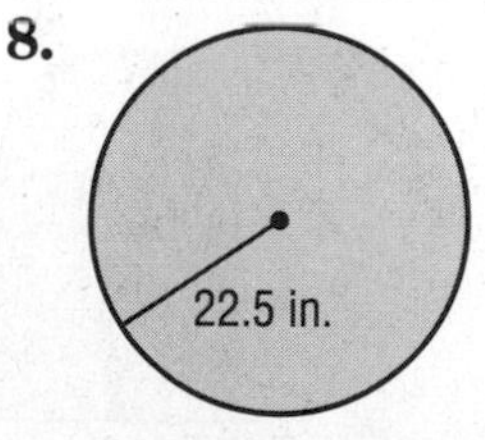

10.

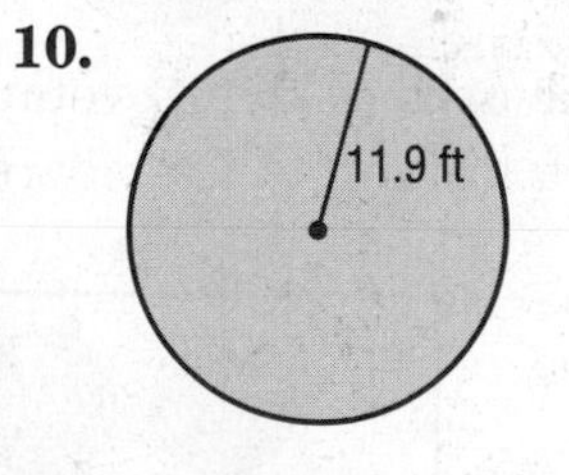

11. radius = 5.7 mm

12. radius = 8.2 ft

13. diameter = $3\frac{1}{4}$ in.

14. diameter = 15.6 cm

15. radius = 1.1 in.

16. diameter = $12\frac{3}{4}$ yd

Lesson 11-4

NAME ______________________________ DATE ______________ PERIOD ______

11-5 Skills Practice

Problem-Solving Investigation: Solve a Simpler Problem

Solve a simpler problem to solve.

1. **POOL** Find the area of the sidewalk around the pool shown below.

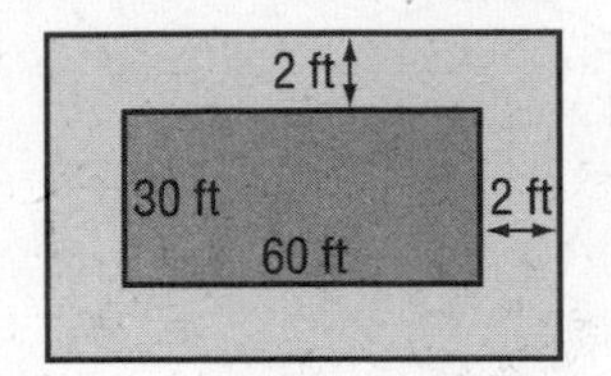

2. **GEOMETRY** Find the area of the shape shown.

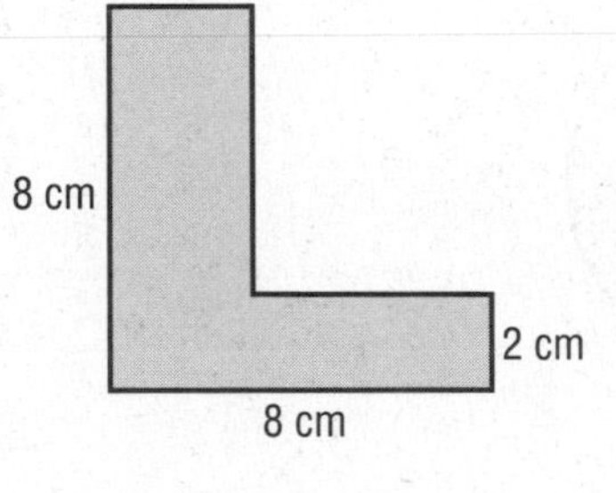

3. **POPULATION** The population of Ghostown, USA is decreasing at a rate of 3 people per year. If there are currently 831 people living in the town, when will the town be deserted?

4. **STAINED GLASS** Find the area of the stained glass window shown below. Use 3.14 for π. Round to the nearest hundredth if necessary.

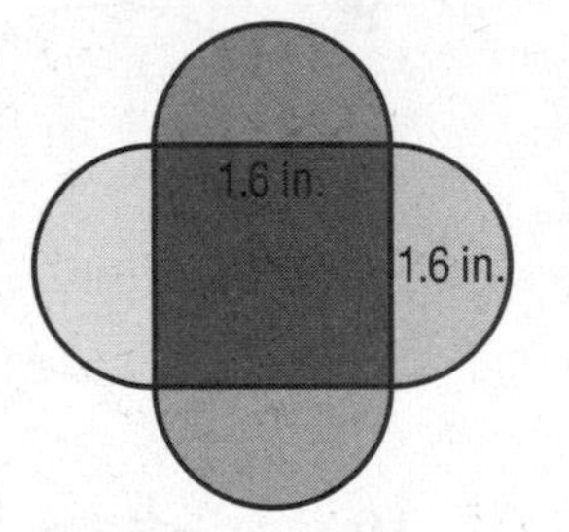

5. **STOVETOPS** What is the area of the stovetop shown, not including the burners? Use 3.14 for π. Round to the nearest hundredth if necessary.

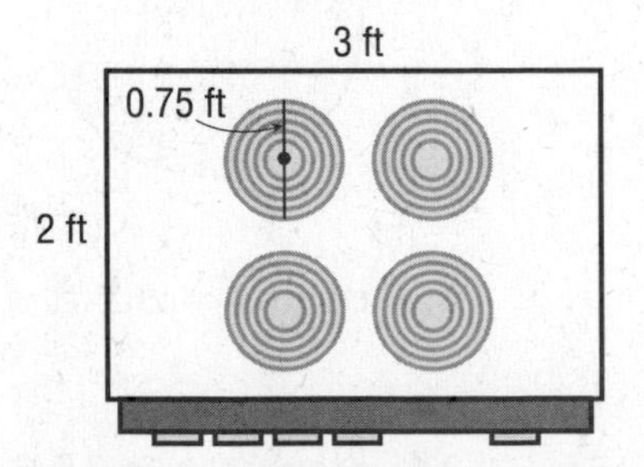

6. **POOLS** Water is being added at a rate of 50 gallons per minute to a pool. How long will it take until the 10,000 gallon pool is full?

11-6 Skills Practice

Area of Composite Figures

Find the area of each figure. Use 3.14 for π. Round to the nearest tenth if necessary.

1.

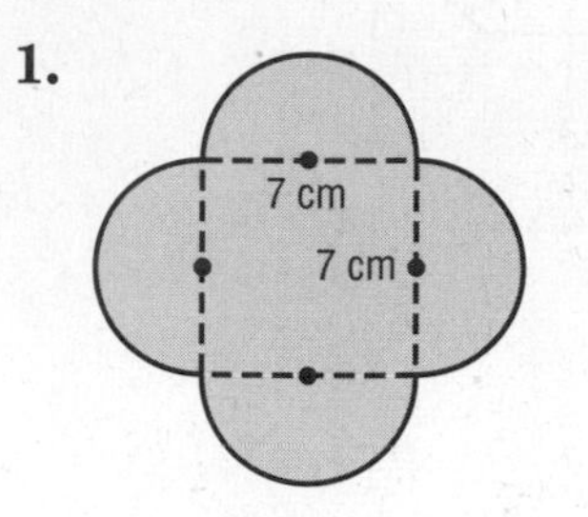

2.

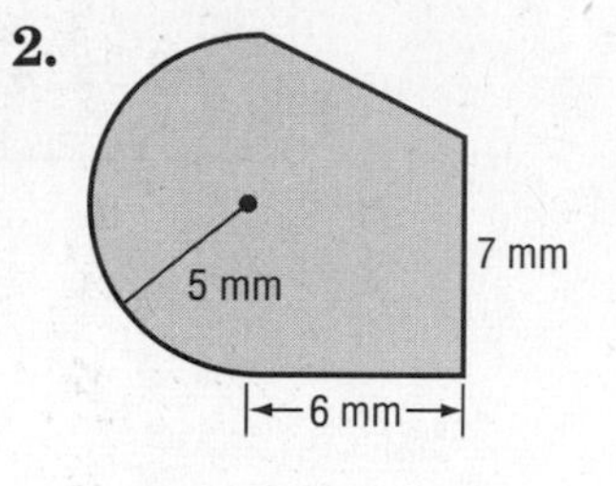

3.

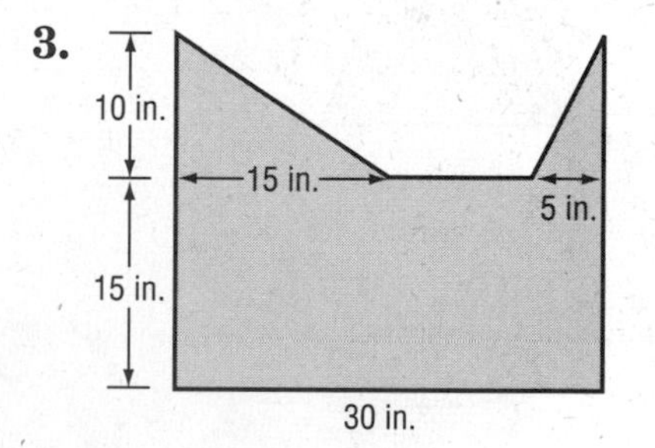

4.

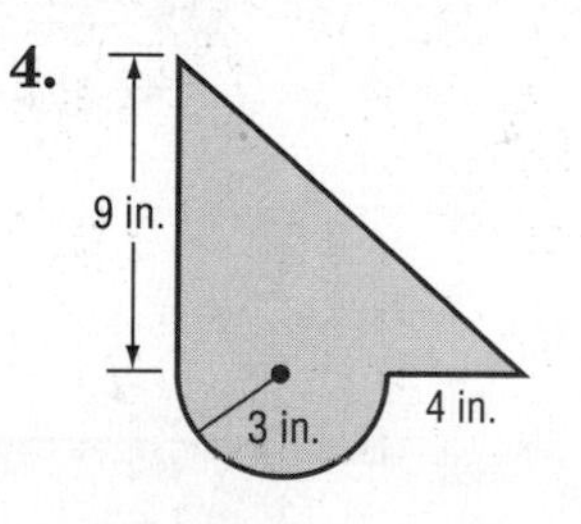

5.

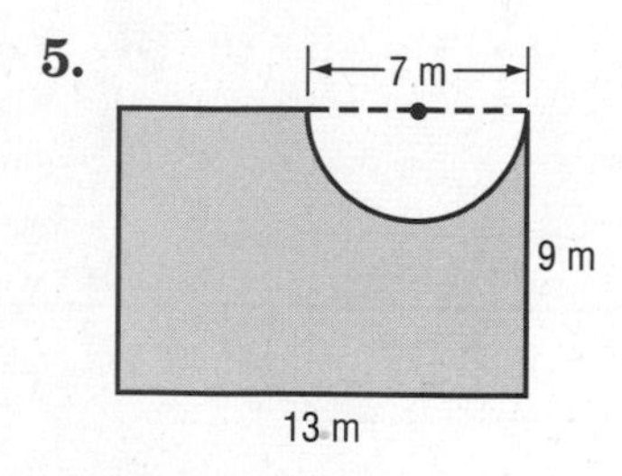

6.

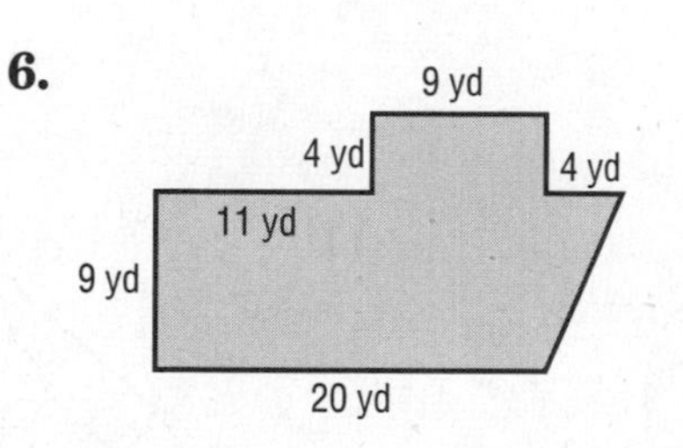

7.

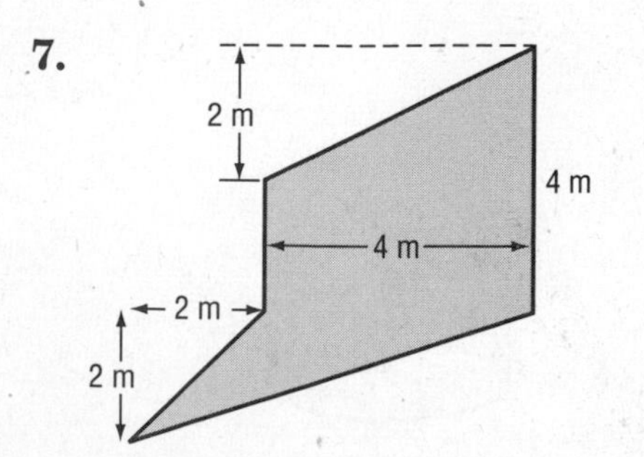

8.

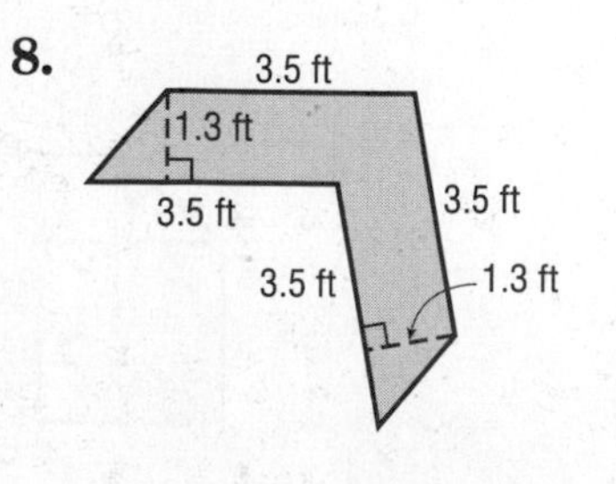

11-7 Skills Practice

Three-Dimensional Figures

For each figure, identify the shape of the base(s). Then classify the figure.

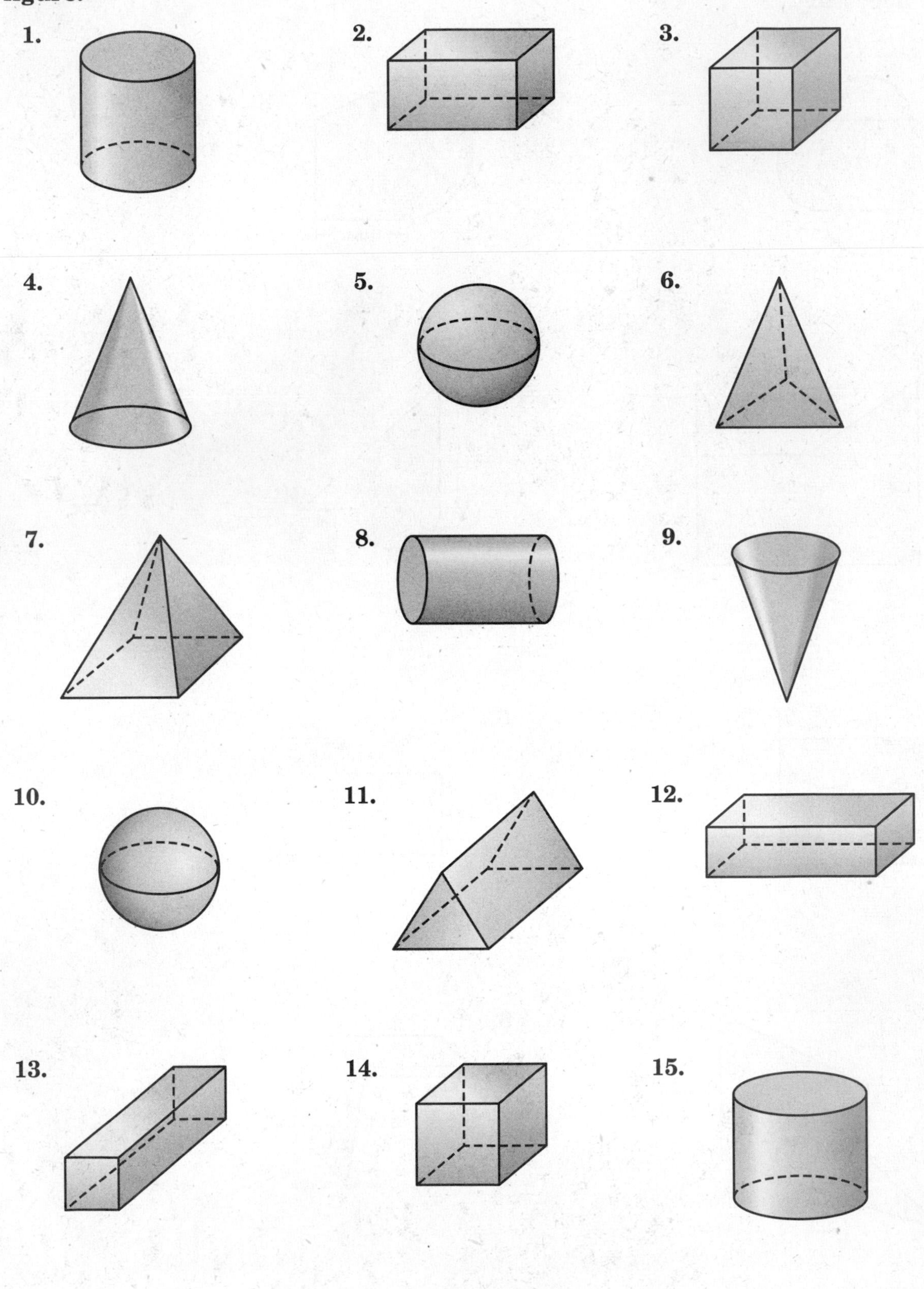

NAME ______________________________ DATE ______________ PERIOD ______

11-8 Skills Practice

Drawing Three-Dimensional Figures

Draw a top, a side, and a front view of each solid.

1.

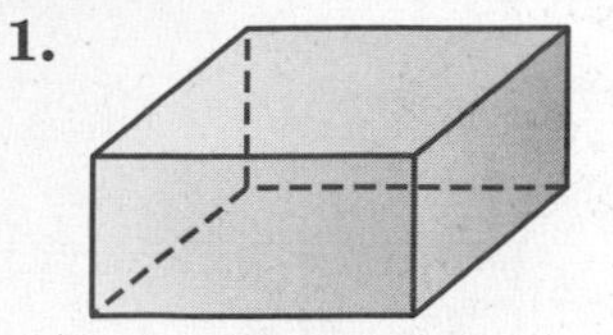

2.

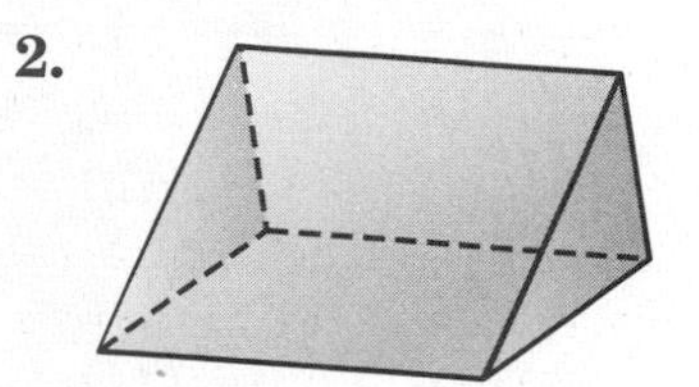

3.

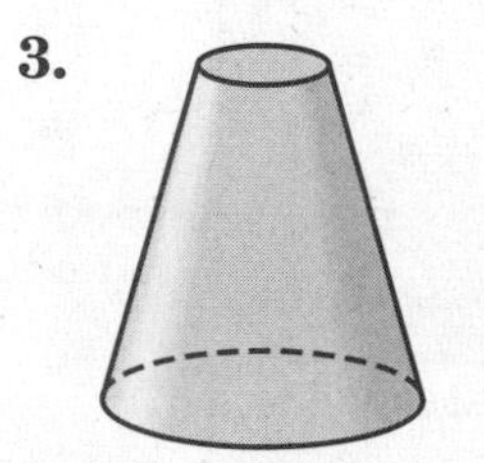

Draw a corner view of each three-dimensional figure whose top, side, and front views shown. Use isometric dot paper.

4.

5.

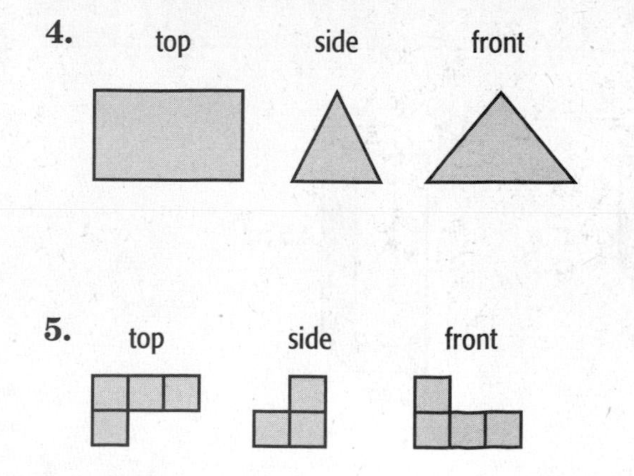

NAME  DATE ______________ PERIOD _____

11-9 Skills Practice

Volume of Prisms

Find the volume of each rectangular prism. Round to the nearest tenth if necessary.

1. **2.** **3.**

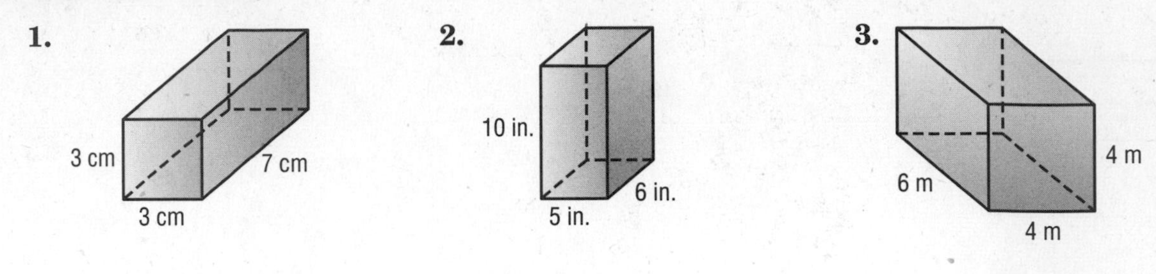

4.

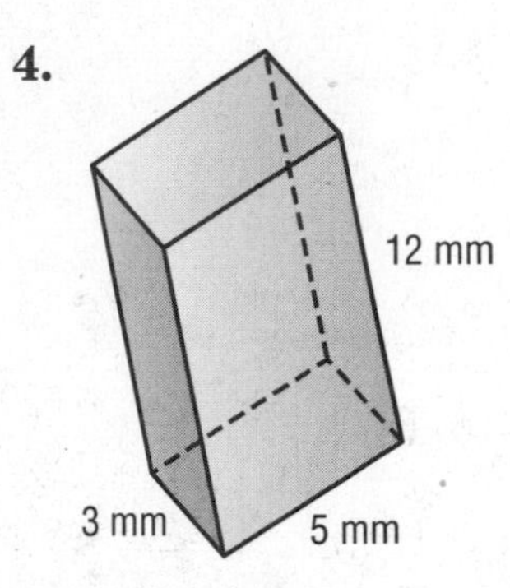

5.

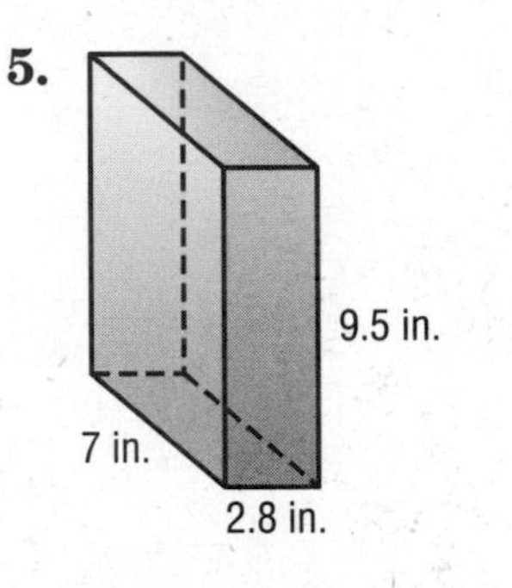

6.

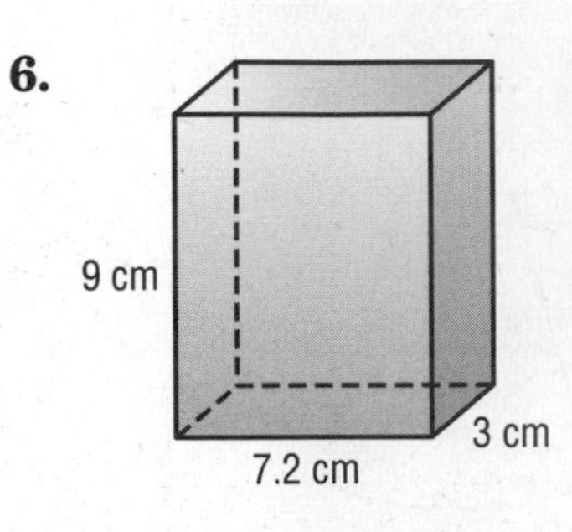

7.

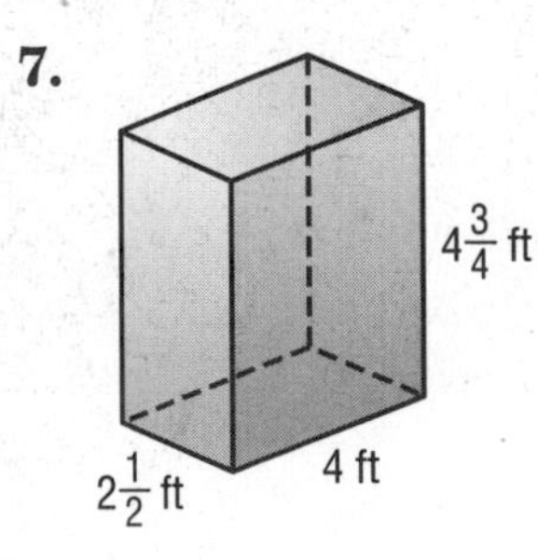

8.

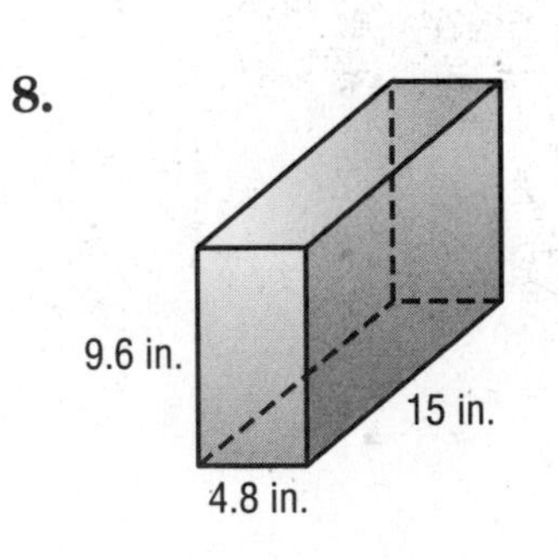

9.

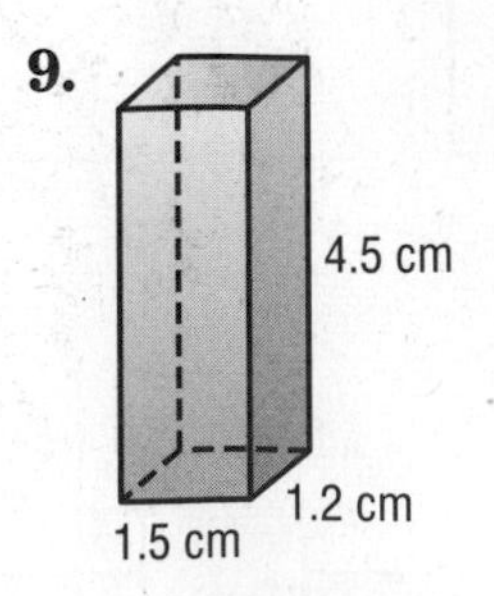

NAME ______________________ DATE ____________ PERIOD _____

11-10 Skills Practice

Volume of Cylinders

Find the volume of each cylinder. Use 3.14 for π. Round to the nearest tenth.

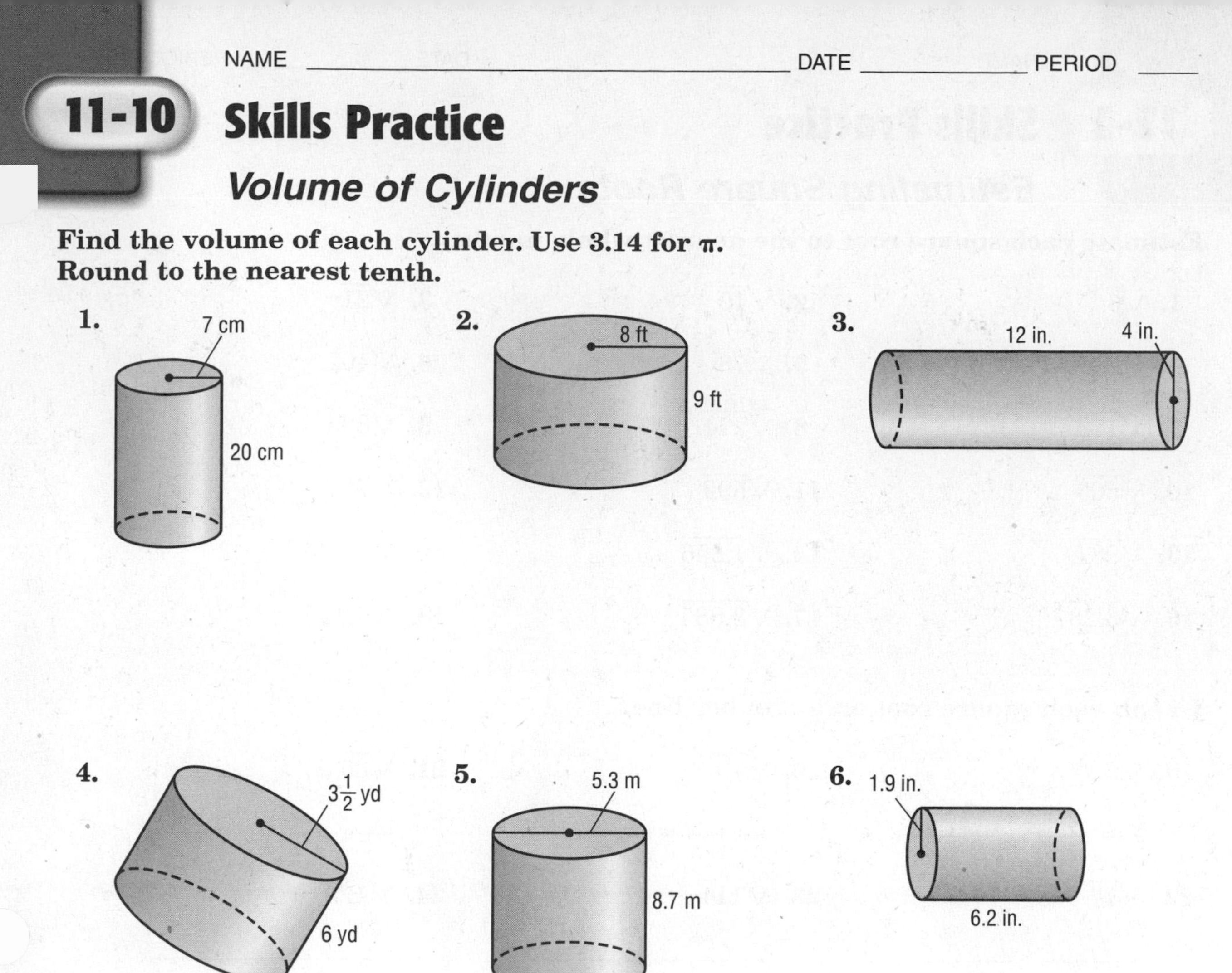

7. radius = 8.8 cm
height = 4.7 cm

8. radius = 4 ft
height = $2\frac{1}{2}$ ft

9. diameter = 10 mm
height = 4 mm

10. diameter = 7.1 in.
height = 1 in.

Lesson 11-10

NAME ______________________________ DATE ____________ PERIOD _____

12-1 Skills Practice

Estimating Square Roots

Estimate each square root to the nearest whole number.

1. $\sqrt{5}$

2. $\sqrt{10}$

3. $\sqrt{21}$

4. $\sqrt{28}$

5. $\sqrt{78}$

6. $\sqrt{102}$

7. $\sqrt{179}$

8. $\sqrt{274}$

9. $\sqrt{303}$

10. $\sqrt{563}$

11. $\sqrt{592}$

12. $\sqrt{755}$

13. $\sqrt{981}$

14. $\sqrt{1,356}$

15. $\sqrt{1,688}$

16. $\sqrt{3,287}$

17. $\sqrt{3,985}$

18. $\sqrt{4,125}$

Graph each square root on a number line.

19. $\sqrt{6}$

3 2 1 0 1 2 3 4 5

20. $\sqrt{19}$

2 1 0 1 2 3 4 5 6

21. $\sqrt{30}$

1 2 3 4 5 6 7 8 9

22. $\sqrt{77}$

6 7 8 9 10 11 12 13 14

23. $\sqrt{114}$

3 4 5 6 7 8 9 10 11

24. $\sqrt{125}$

6 7 8 9 10 11 12 13 14

25. $\sqrt{149}$

6 7 8 9 10 11 12 13 14

26. $\sqrt{182}$

6 7 8 9 10 11 12 13 14

27. $\sqrt{212}$

13 14 15 16 17 18 19 20 21

28. $\sqrt{436}$

13 14 15 16 17 18 19 20 21

29. $\sqrt{621}$

21 22 23 24 25 26 27 28 29

30. $\sqrt{853}$

25 26 27 28 29 30 31 32 33

31. $\sqrt{918}$

25 26 27 28 29 30 31 32 33

32. $\sqrt{1,004}$

25 26 27 28 29 30 31 32 33

33. $\sqrt{1,270}$

30 31 32 33 34 35 36 37 38

34. $\sqrt{5,438}$

70 71 72 73 74 75 76 77 78

35. $\sqrt{4,215}$

60 61 62 63 64 65 66 67 68

36. $\sqrt{5,786}$

70 71 72 73 74 75 76 77 78

37. Order $\frac{25}{7}$, 4.91, and $\sqrt{23}$ from least to greatest.

38. Graph $\sqrt{42}$ and $\sqrt{62}$ on the same number line.

0 1 2 3 4 5 6 7 8

12-2 Skills Practice

The Pythagorean Theorem

Find the missing measure of each right triangle. Round to the nearest tenth if necessary.

1.

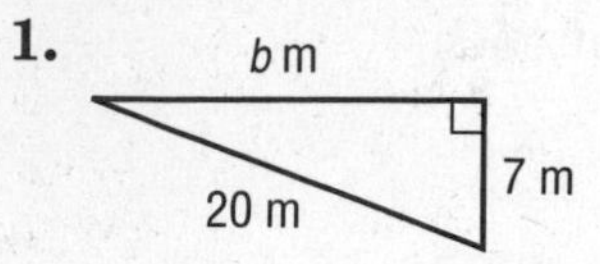

2.

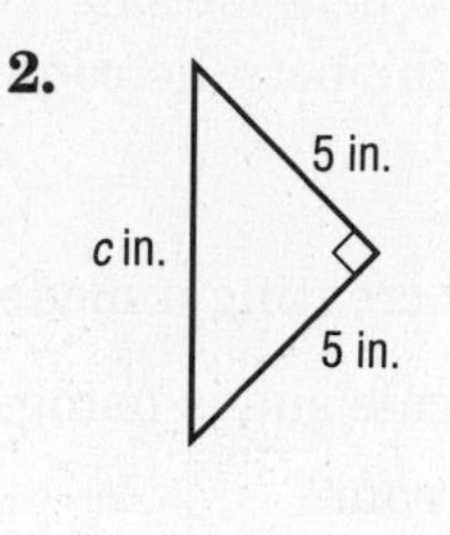

3.

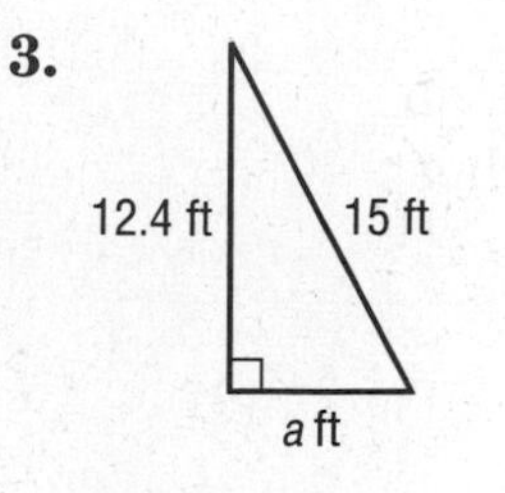

4.

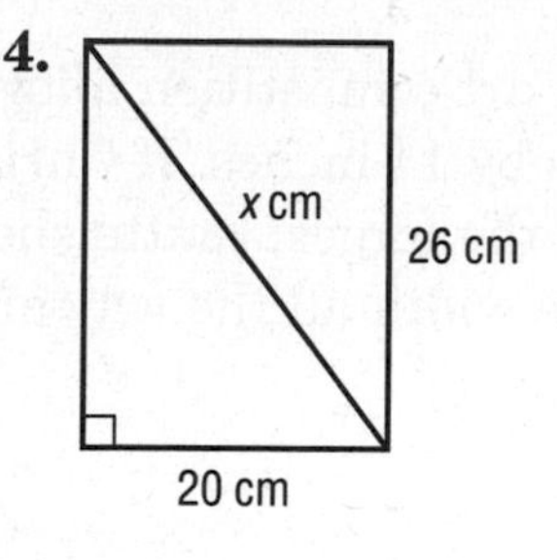

5.

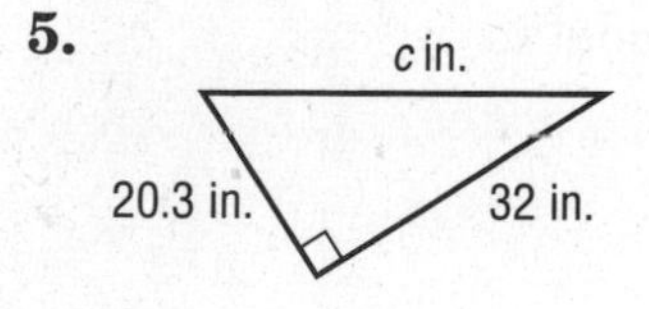

6.

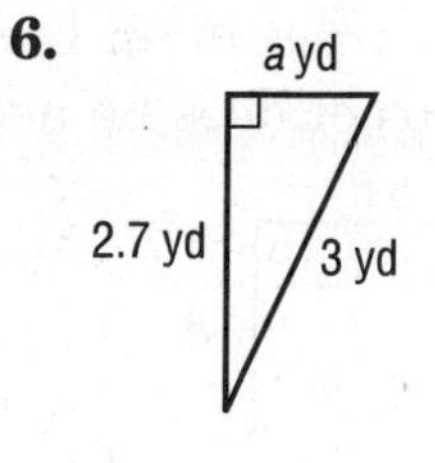

7.

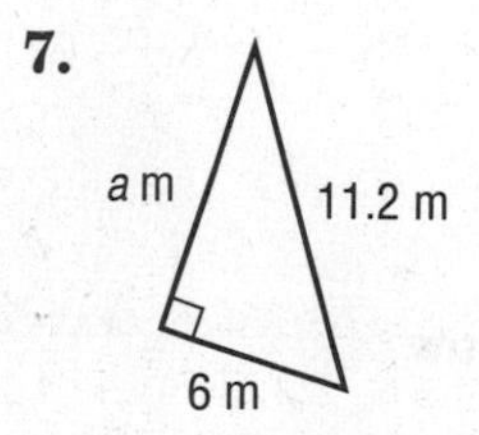

8.

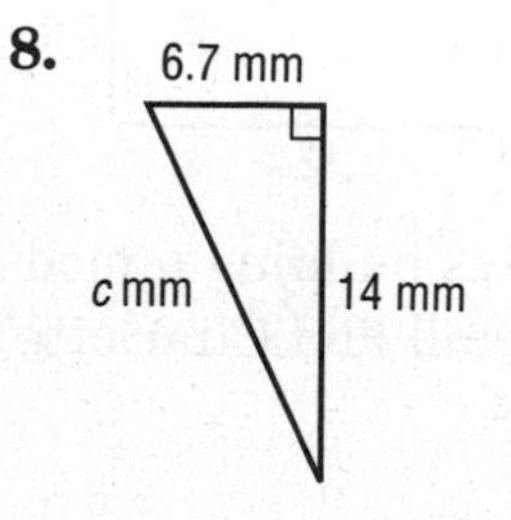

9. $a = 15$ cm, $b = 20$ cm

10. $a = 2$ yd, $b = 12$ yd

11. $a = 13$ in., $c = 16.5$ in.

12. $b = 8$ mm, $c = 17$ mm

13. $a = 1.3$ ft, $b = 4.6$ ft

14. $a = 14.7$ m, $c = 23$ m

15. $a = 10$ ft, $b = 24$ ft

16. $b = 8$ in., $c = 9$ in.

17. $b = 9$ cm, $c = 12$ cm

18. $a = 4.5$ mm, $c = 7.5$ mm

NAME ______________________ DATE ____________ PERIOD _____

12-3 Skills Practice

Problem-Solving Investigation: Make a Model

Make a model to solve.

1. **PETS** Jack's Pet Store has 5 pets for sale. Some are birds and some are dogs. When Jack looks at the pets, he counts 18 legs. How many of each type of pet are there?

2. **INTERIOR DESIGN** JoAnn is creating a model of a living room. The room is 20 feet by 20 feet. If the scale she is using is 1 foot = $\frac{1}{2}$ inch, what are the dimensions of her model room?

3. **ART COMPETITION** An art competition allows for submitted work to be no larger than 11 inches by 14 inches. If Christene's photograph is 8 inches by 10 inches, what is the largest matte she can use if she wants the border to be the same width all the way around the photo?

4. **FLOORING** James is laying carpet in an L-shaped room whose model is shown below. How much carpet does he need?

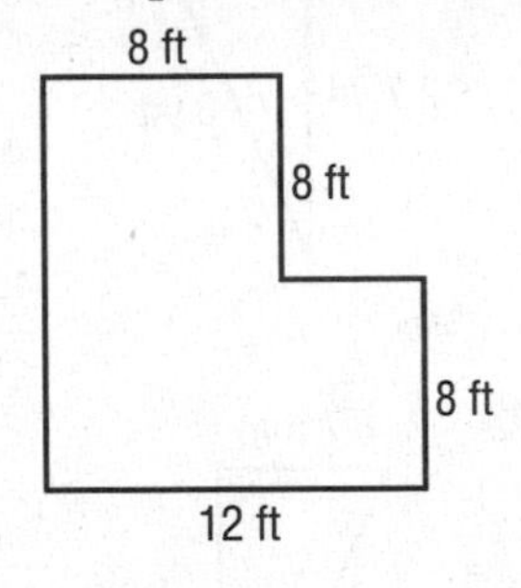

5. **SCALE MODEL** Charlotte is building a model of the Eiffel Tower. If the actual tower is 986 feet tall and Charlotte's scale is 1 inch = 10 feet, how tall is her model?

6. **SCIENCE FAIR** Audrey wants to make a poster that is folded into three sections for her science fair project. The length of the poster is 36 inches. If she wants the middle section to be twice the length of the side sections and she wants the two side sections to be equal, what should be the length of the middle section?

12-4 Skills Practice

Surface Area of Rectangular Prisms

Find the surface area of each rectangular prism. Round to the nearest tenth if necessary.

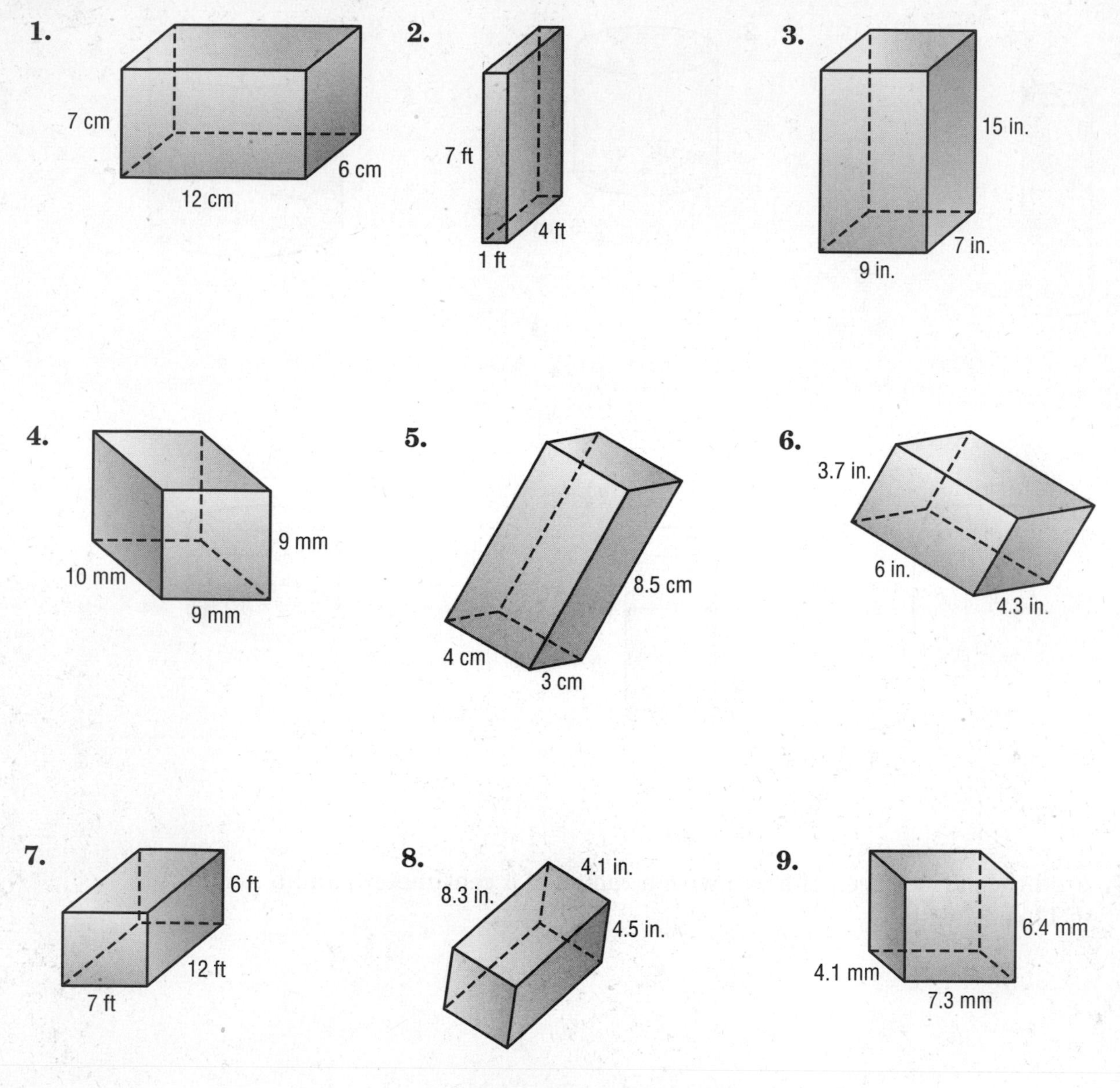

10. A cube has a surface area of 126 square feet. What is the area of one face?

11. Find the surface area of a rectangular prism that has a length of 8 inches, a width of 3 inches, and a height of 6 inches.

NAME ______________________ DATE ____________ PERIOD _____

12-5 Skills Practice

Surface Area of Cylinders

Find the surface area of each cylinder. Use 3.14 for π. Round to the nearest tenth.

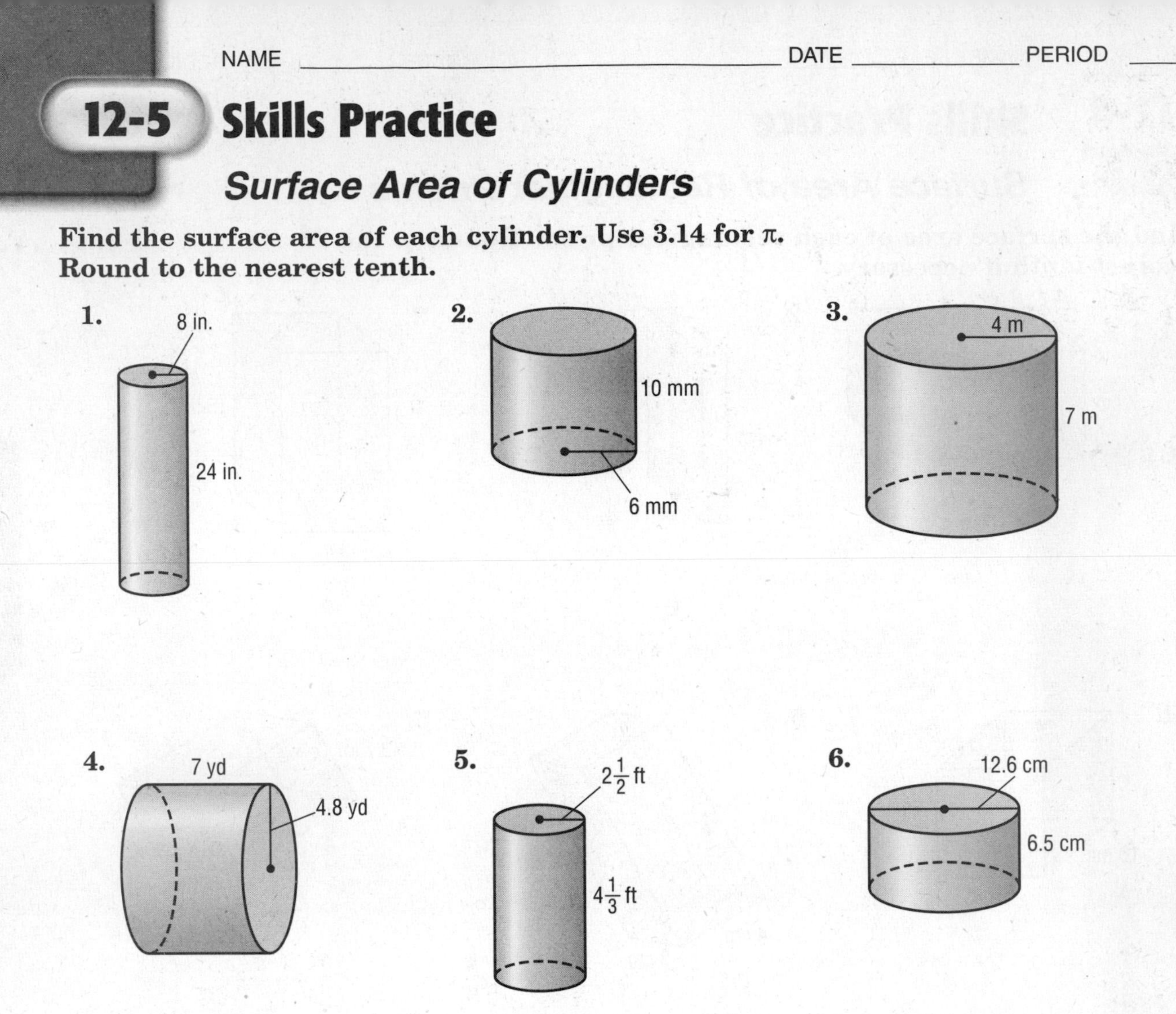

7. Find the surface area of a can with a radius of 4 centimeters and a height of 11 centimeters.

8. Find the surface area of the outside of a cylindrical barrel with a diameter of 10 inches and a height of 12 inches.

9. Find the area of the curved surface of a D battery with a diameter of 3.2 centimeters and a height of 5.6 centimeters.